VOLUME 2
RANDOM COUNTS IN SCIENTIFIC WORK

THE PENN STATE STATISTICS SERIES

An International Series in Statistics and Applications

General Editor: G. P. Patil, Professor of Statistics
The Pennsylvania State University

This new series brings volumes of importance to statisticians and scientists in related disciplines.
The series is organized in three general categories:

I. Monographs on Statistical Distributions and Applications

II. Monographs on Related Topics in Statistics

III. Edited multi-author volumes on subjects of contemporary interest in statistics

RANDOM COUNTS IN SCIENTIFIC WORK
Expanded from the Proceedings of the
Biometric Society Symposium
Dallas, Texas, December, 1968
(American Association for the
Advancement of Science Conference)

VOLUME 2
RANDOM COUNTS IN BIOMEDICAL AND SOCIAL SCIENCES

G. P. PATIL, Editor
Department of Statistics
The Pennsylvania State University

The Pennsylvania State University Press
University Park and London

Library of Congress Catalogue Card Number 73-114351

Standard Book Number 271-00115-1

Printed in the United States of America

The Pennsylvania State University Press
University Park, Pennsylvania 16802

The Pennsylvania State University Press, Ltd.
London W. 1

CONTENTS

PREFACE

As Professor C. R. Rao has put it: "Endowed with a wide and rapidly expanding literature, both in diverse fields of applied science and statistics, the subject of random counts in scientific work or discrete distributions and their applications offers a fascinating area of study and research." It is of course conceivable that older and more familiar theories of statistical inference may be specifically needed or used to analyze exclusively the data on random counts. However, it is evident that new and novel mathematical and statistical approaches and procedures pertaining to special types of counted data related to individual scientific problems are long overdue.

These new procedures may be evolved in two ways. One method is to investigate in detail individual discrete distributions for their mathematical and statistical properties and for their natural applications. The second approach is to concentrate on a scientific problem that generates or requires counted data and then create and coordinate appropriate discrete models and methods so as to arrive at a meaningful solution of the scientific problem.

Random counts thus has much to offer for the advancement of the basic and applied sciences, as well as for the advancement of statistics itself.

From this premise and the multidisciplinary background of the annual conferences of the American Association for the Advancement of Science, I had felt for some time that the Biometric Society and the Statistics Section of the AAAS would provide the proper atmosphere for a much-needed program on the subject of random counts in scientific work.

During the time that I served on the program committee of the Biometric Society, and while I was the program chairman for 1968 meetings, I had several instructive and encouraging discussions with the immediate past program chairmen D. S. Robson and E. C. Pielou, with H. O. Hartley, Biometric Society ENAR President, and with C. I. Bliss, the Vice-President of the Statistics Section of the AAAS. A symposium on the subject of random counts was planned and organized at the Dallas Conference of the Biometric Society in December of 1968; twenty participants were on the program.

The symposium was a great success in that there was a very lively exchange and interaction between scientists from different fields and statisticians interested in the analysis of random counts. Although there seemed to be divergence of levels and diversity in the presentations, this turned out to be more of an asset than a liability.

Because of the quality and value of the presented papers, as well as widespread interest in the topic, the participants and audience felt that the proceedings should be published. It also seemed appropriate to invite additional papers from specialists who were not present in Dallas. Twenty-two papers were invited and organized in this category.

For purposes of convenience, the final total of 42 papers has been divided into three discrete subject-matter groups, published in a like number of volumes under the general title *Random Counts in Scientific Work*, treating random counts in (1) models and structures, (2) biomedical and social sciences, and (3) physical sciences, geoscience, and business. The present volume, *Random Counts in Biomedical and Social Sciences*, represents the second of the three volumes. It is expected that more volumes will be added to this series as progress and new results in the field of random counts in scientific work come about. Any comments and suggestions from readers would be most welcome.

University Park, Pennsylvania — G. P. Patil
April 9, 1970

VOLUME 2
RANDOM COUNTS IN SCIENTIFIC WORK

STATISTICAL CONTROL IN HAEMATOLOGY

H. O. LANCASTER
Department of Mathematical Statistics
The University of Sydney

SUMMARY

The Poisson and some alternative models of the distribution of red cells in the haemocytometer are re-examined and some remarks are made on the "crowding effect", under which the variance is diminished below the theoretical Poisson value. The X^2 approximation for the test of homogeneity between replicate counts has been found to be adequate in various sampling experiments and by complete enumeration of cases with certain given sample totals and number of replicates.

1. INTRODUCTION

Historical notes on the application of statistical methods to the quality control of such counting experiments as occur in bacteriology and haematology may be found in [5], [10], and [20]; and the reader is referred to them for detailed remarks on theory. Plum [17] gives a good account of the technical methods and provides excellent observational material. We may pass over the earlier work and the various independent discoveries of the Poisson distribution (see [9]) to the counts of yeast cells in a haemocytometer by "Student" [21], who found that the Poisson distribution gave a good fit to the distributions of his complete counts of the 400 haemocytometer square areas, when he used the large sample methods for the application of χ^2. In haematology the method would be time consuming and, moreover, would not give the necessary quality control of counts as they are carried out in clinical laboratories. Nor would the large sample methods be practicable in bacteriology, where the problem is often to test the consistency of the counts on "parallel" bacterial plates, that is, replicate plates sown with the same volume of dilute suspension of bacteria. Fisher, Thornton, and MacKenzie [8] devised a small-sample method appropriate to the study of such counts. These methods have been made familiar to bacteriologists by their inclusion in the successive editions of R. A. Fisher's Statistical Methods.

Although "Student's" work had been favorably reviewed by a leading haematologist, little use had been made of the methods of statistical control in haematology until the 1930's, even though sampling conditions can easily be made to approximate to the ideal and one might have expected the methods to become standard. However, errors of interpretation soon began to arise when statistical methods were introduced; for example, textbook authors were giving absurd criteria for the consistency or agreement between parallel counts of red cells as carried out under the ordinary laboratory conditions. J. Berkson and his colleagues at the Mayo Clinic warmly criticized these criteria and carried out some fine investigations on the technical side. They were able to check the counts of technicians and interns against photographs of the haemocytometer chamber. Berkson, Magath and Hurn [4] noticed, in particular, a regular departure from the values expected under Poisson theory, the "crowding" effect, which will be discussed later in this paper.

The later sections of this paper deal with the Poisson model for counting experiments, the sizes of the tests, the departures of the observations from the theoretical, and various applications of the tests.

2. THE POISSON MODEL

The objects to be counted (in haematology, the red cells in a diluting fluid) are of small volume and can be regarded as points (negligible in size). This is a reasonable approximation. In fact, the red cells of a normal human subject in the commonly used 1-in-200 dilution occupy about 0.2% of the volume. We may consider now a unit volume of a prepared dilution and imagine it filled by a Poisson process with parameter λ, so that the expectation of the number of cells appearing in a unit volume is λ. It is a well-known theory (see [9]) that the probability of obtaining x cells in a given unit volume is given by

$$P(X=x) = \exp(-\lambda)\lambda^x/x! = p_x, \text{ say.} \tag{2.1}$$

When λ is known, the test of the Poisson process is to carry out n independent observations from the same dilution (or same λ) and classify the counts by the values of x, so that a_x counts of x are observed. Student [21] used the Pearson χ^2, of the form

$$Q = \sum_x (a_x - np_x^*)^2/(np_x^*). \tag{2.2}$$

Some pooling of classes will always be necessary in the upper tail of the distribution and, if the mean counts are greater than about 5, will also be necessary at the lower tail of the distribution of the observed counts. In (2.2) p_x^* is to be interpreted as p_x for the classes for which no pooling has been used and p_x^* is a sum of probabilities in the classes pooled. Usually λ is not known, in which case an estimate, $\hat{\lambda}$, can be used:

$$n\hat{\lambda} = \sum_{x=0}^{\infty} xa_x \tag{2.3}$$

is the sufficient maximum likelihood estimate. It may be noted that, since pooling must be used, the estimator of (2.3) may not be appropriate. In this case a mean value may be assigned to the pooled classes, in the upper tail, say, γ_0. $\hat{\lambda}$, the first estimate of λ, can then be computed, and then a value γ_1 can be assigned to the upper class being the expected value for this class under the assumption that the distribution is Poisson with parameter, $\hat{\lambda}_1$. This procedure can be continued with (2.3) replaced by

$$n\hat{\lambda}_j = \sum_{x=0}^{k=1} xa_x + \gamma_{j-1}a_k^* , \tag{2.4}$$

where a_k^* is the number of observations in the pooled class. This procedure can be continued until successive values of γ_j and hence of $\hat{\lambda}_j$ are sufficiently close. The Pearson test function for goodness-of-fit is now (2.2) distributed with k - 1 degrees of freedom, or with degrees of freedom two less than the number of classes after pooling; and the process of fitting has set equal to zero the linear comparison of the class totals represented by (2.4). The iteration above would be appropriate if pooling was carried out in more than one class.

Some care must be taken if $\hat{\lambda}$ is small and there are few classes, say $\{0\}$, $\{1\}$, and $\{x:x>1\}$, or $\{0\}$, $\{1\}$, $\{2\}$ and $\{x:x>2\}$, for the test is quite insensitive to departures from the Poisson hypothesis in these cases. Even with larger values of $\hat{\lambda}$, it is insensitive to certain departures (as in the sticking together of white cells in blood counting) if too much pooling is carried out in the upper tail of the distribution, as Berkson [3] has pointed out. With low means of the order of $\lambda = 1$, it is preferable to use a combinatorial test such as that due to Fisher [7].

The Poisson distribution can be characterized by the property that, if $X_1, X_2, \ldots, X_n$ are independent Poisson variables with parameters $\lambda q_1, \lambda q_2, \ldots \lambda q_n$, then the joint distribution of $\{X_i\}$ conditional on $\Sigma X_i = N$ is a multinomial with index N and class probabilities equal to $\{q_i\}$, as in Moran [16]. Indeed, if $\{x_1, x_2, \ldots, x_n\}$ is a composition of N with zeroes permitted so that $\Sigma x_i = N$, we have

$$P(\Sigma X_i = N) = \Sigma \prod_1^n P(X_i = x_i), \tag{2.5}$$

with summation over the compositions of N. But from (2.5) it follows that

$$P(\Sigma X_i=N) = \exp(-\lambda)\lambda^N \Sigma q_1^{x_1} q_2^{x_2} \ldots q_n^{x_n} (x_1!x_2!\ldots x_n!)^{-1} , \tag{2.6}$$

but the result of the summation is $(N!)^{-1}$, since the terms summed are those of the multinomial $(q_1+q_2+\ldots+q_n)^N$ divided by $N!$. (2.6) thus yields $P(\Sigma X_i=N) = \exp(-\lambda)\lambda^N/N!$. Now we obtain for any set $\{x_i\}$, subject to $\Sigma x_i = N$,

$$P(\{X_i\} = \{x_i\}|\Sigma X_i=N) = P(\{X_i\} = \{x_i\}|\{\lambda q_i\}, \text{Poisson})/P(\Sigma X_i=N)$$

$$= N! \prod_1^n [q_i^{x_i}/x_i!], \tag{2.7}$$

the multinomial expression. In the case of special interest to us each of the Poisson parameters is the same so that $q_i = n^{-1}$, and (2.7) simplifies to

$$P(\{X_i\}|\Sigma X_i=N) = n^{-N}N!/\prod_1^n x_i! \tag{2.8}$$

and an appropriate test of significance is

$$Q = \Sigma(X_i-Nn^{-1})^2/Nn^{-1}$$

$$= n\Sigma X_i^2/N - N \tag{2.9}$$

$$= n\Sigma X_i^2/\Sigma X_i - \Sigma X_i,$$

as introduced by [8]. We defer considerations of the use of the test to a later section.

3. THE SIZES OF THE TESTS

There has been much discussion as to the accuracy of the χ^2 approximation when some expected numbers are not large. The main difficulty (which applies to any significance test when the variable has a discrete distribution) is that for a fixed sample size N there may be few configurations associated with non-negligible probabilities (relative frequencies). Thus with two classes in the counting problem we have X_1 and X_2, the duplicate counts, and $N = X_1 + X_2$. χ^2 is $4(X_1 - \frac{1}{2}N)^2/N$. The number of distinct partitions of N accounting for, say,

99% of the probability is the integral part of $\frac{1}{2}N^{\frac{1}{2}}t$, where t is the 0.01 point of the normal distribution, 2.56. Thus for N = 25 there are only 6 configurations (in this case partitions of N) contributing 99% of the total probability. With three or more counts we may write

$$Q = \Sigma(X_i - \bar{X})^2/\bar{X} - \Sigma X_i$$
$$= n\Sigma X_i^2/\Sigma X_i - \Sigma X_i, \tag{3.1}$$

and the number of distinct values to be considered is the number of partitions of $N \equiv \Sigma X_i$ into three parts with zeroes allowed. Not all these distinct partitions yield distinct values of Q since for a fixed N, Σx_i^2 can be equal to Σy_i^2, where $\{x_i\}$ and $\{y_i\}$ are distinct partitions of N. Once again, it can be shown that for a fixed N, there are relatively few distinct values of Q and so no close approximation to size is possible. However, Lancaster and Brown [15] showed by a complete enumeration that approximations of the sizes of the tests in the discrete distributions to the nominal sizes of the theoretical distributions were good if the sample size, $N = \Sigma X_i$, had the Poisson distribution. In practice, the counts observed come from mixtures of mixtures. In other words, N is a random variable with parameter $n\lambda$, and λ itself is a random variable, for the counts being examined are generated by different patients in the blood counting type of experiment or the bacterial counts are being obtained from different soils, water supplies, and so on.

There have been a number of investigations of the adequacy of the χ^2 approximation, and most authors have expressed themselves as agreeably surprised at the results of their computations. References appropriate to this Section are [2], [6], [10]-[15], [18], [19], and [24].

It may be noted that in the sampling experiments of el Shanawany [6], who was considering, among others, sets of 15 counts with Poisson parameter equal to 2, authors have claimed that the good results are due to the many different values of the Q of our (2.9), above, for the fixed N. This can easily be shown to be false by a complete enumeration of the partitions of, say, 30 into 15 parts with zeroes allowed, which have a non-negligible probability. The favorable results came about because under the sampling conditions the total N of the counts, the index of the multinomial, was itself a random variable; and thus Sukhatme's N had the distribution of a mixture.

We conclude that the sizes of the tests satisfactorily approximate the theoretical.

4. THE CROWDING EFFECT

In counts of the red cells of healthy subjects made under standard conditions, there appears a "crowding effect," first noticed by Berkson, Magath, and Hurn [4], who found that the sample variance had an expectation of approximately 0.82 λ, where λ, the expected number of cells on a small square of the haemocytometer, was about 7 or 8. We now consider this effect in detail.

The mathematical model of a red cell count makes assumptions, according to Lancaster [10], as follows:

(a) complete mixing has been attained;

(b) there is no cohesion between cells; and

(c) the fact of one cell falling onto a certain square of the haemocytometer does not affect the changes of another doing so.

However, Lancaster (also in [10]) showed that the variance, v, could be related to the observed mean, m, in counts over the 400 small squares by an equation of the form

$$v = m - bm^2. \tag{4.1}$$

For his own data b was 0.020534, and for the data of Berkson, Magath, and Hurn [4] it was 0.026324. All observers are agreed on the reality of the diminished variance, or crowding, effect, although there are differences on how this should be represented in the mathematical model.

On the physical side, there are two processes to be considered: first, a mixing in the suspension, and second, the settling of the cells onto the floor of the haemocytometer well. As to the first, complete mixing is practicable; cohesion between the cells does not occur, but the cells possess a finite volume, and there is some interference between them as a result. However, the cells only occupy 0.025λ% of the total volume, where λ is the expectation per small square (λ = 8, approximately, for healthy subjects). It is probable that this effect is negligible and could only be detected by experimental means with great difficulty. We may assume provisionally that the number of cells in the fluid immediately above a small square area is effectively a Poisson variable. In the second process, the cells settle out from the suspension onto the floor of the haemocytometer well. Here, the dimensions of the red cell are not negligible. The side of the small square is 1/20 mm or 50μ. The diameter of the red cells is about 8μ) and they can be regarded as a disc rather more than 1μ in thickness. The explanation of the crowding effect given by Lancaster [10] is as follows. It may be assumed that, when one red cell has fallen onto a square, if a second falls directly onto it, the second cell will slip over the first until the two discs are just in contact; in so slipping, the second disc may pass over the boundary of the square into another square. The consequences of this model have been worked out by a geometrical argument and imply a diminution of the variance of the "right order." It is evident that the probability of a square losing a cell will be greater, the larger the number of cells to fall on it.

An approximate theory can be given using the last remark and the empirical result (4.1). Since the cells are displaced from one square to a neighboring square, there will be correlations between neighboring cells; and the greatest of these correlations will be between two squares sharing a common boundary, the common side. The correlation between squares sharing an apex is assumed to be negligible. The variance of the count on a square area containing k^2 small squares can now be computed. There are k^2 variances, σ^2, say, of the individual squares to be considered and $2k(k-1)$ pairs of covariances, $\rho\sigma^2$, since this is the number of pairs of small squares with common boundaries. By common statistical theory, we may write for the variance

$$V(k^2) = \{k^2+4k(k-1)\rho\}\sigma^2. \tag{4.2}$$

On the other hand, if k is large, there will be a central square area of dimensions about $(k-1)^2$, from which there will be no loss of cells to other large square areas and a border area from which cells may be lost to other large square areas and similarly to which gains of cells may be made. The central square area is of dimension k^2, and the marginal area is of dimension k, so that we may write

$$V(k^2)/(k^2\lambda) = 1, \tag{4.3}$$

if k is large, since the number of cells in suspension above the central area will have a Poisson distribution and all these cells will be included in the count on the k^2 squares. Comparing (4.2) and (4.3), we obtain

$$\lambda = (1+4\rho)\sigma^2, \quad \rho = (\lambda/\sigma^2-1)/4. \tag{4.4}$$

Now that ρ has been obtained, its value can be substituted in (4.2) to obtain

$$V(k^2) = \{\lambda k^2-k(\lambda-\sigma^2)\}. \tag{4.5}$$

From (4.5),

$$V(k^2)/(\lambda k^2) = 1 - (\lambda-\sigma^2)/k \tag{4.6}$$

any acceptable theory of the distribution of cells on the haemocytometer must take account of the correlation between the counts on neighboring squares.

Turner and Eadie [22] set up a model as follows: Let it be imagined that each small (haemocytometer) square is divided up into a spaces or stalls, that A is the maximum number of cells that can be accommodated in the whole haemocytometer chamber, and that there are n small squares so that A = an. Then it can be deduced that the number of ways k cells can be distributed over the n squares, subject to $\Sigma k_j = k$, is given by

$$\prod_{j=1}^{n} \binom{a}{k_j} \Big/ \binom{A}{k} \ . \tag{4.7}$$

By a suitable choice of a, the variance of a count can be reduced to the level required. However, in the model the idealized red cell acts as though, when it finds one stall occupied, it can seek a further stall with equal probability over the whole haemocytometer--an unlikely physical behavior. The model of Turner and Eadie [22] thus ignores the possibility of the correlation between neighboring cells and can give no answer to the question of the variance on a set of k × k small squares.

Barton and David [1] discuss the "interesting result" of Turner and Eadie [22]. In fact, the distribution is that of an n × 2 contingency table in which the individual squares have become the rows, now with columns all equal to the number of stalls, a, in a small square; moreover, no notice is taken of the original geometrical positions of the small squares. This paper also fails to handle the problem of combining neighboring cells into groups of $k \times k$.

Although a convenient method of obtaining the variances is available as in the early part of this Section, there is no general theory and there is reason to believe that no neat analytic results are likely.

5. APPLICATION OF THE METHODS

The theory of Sections 3 and 4 shows first that the χ^2 approximation is adequate for testing the null hypothesis in the multinomial in that the expected sizes of the discrete test approximate closely to the continuous and second that the "crowding effect" can be ignored if counts on groups of 4×4 small haemocytometer squares are made.

For the setting up of quality control of counting, the reader is referred to [8], [5], [10], and [20].

REFERENCES

[1] Barton, D. E., and David, F. N. 1959. Haemocytometer counts and occupancy problems. Trab. Estadist. 10:12-18.

[2] Bennett, R. W. 1962. Size of the χ^2 test in the multinomial distribution. Austral. J. Statist. 4:86-8.

[3] Berkson, J. 1938. Some difficulties of interpretation encountered in the application of the chi-square table. J. Amer. Statist. Assoc. 33:526-36.

[4] ———; Magath, T. B.; and Hurn, M. 1935. Laboratory standards in relation to chance fluctuations of the erythrocyte count as estimated with the haemocytometer. J. Amer. Statist. Assoc. 30:414-26.

[5] Eisenhart, C., and Wilson, P. W. 1943. Statistical methods and control in bacteriology. Bact. Rev. 7:57-137.

[6] el Shanawany, M. R. 1936. An illustration of the accuracy of the χ^2 approximation. Biometrika 28:179-87.

[7] Fisher, R. A. 1950. The significance of deviations from expectation in a Poisson series. Biometrics 6:17-24.

[8] ————.; Thornton, H. G.; and MacKenzie, W. A. 1922. The accuracy of the plating method of estimating the density of bacterial populations. Ann. Appl. Biology 9:325-59.

[9] Haight, F. A. 1967. Handbook of the Poisson distribution. John Wiley & Sons, Inc., New York. Pp. xi, 168.

[10] Lancaster, H. O. 1950. Statistical control in haematology. J. Hyg., Camb. 48:402-17.

[11] ————. 1952. Statistical control of counting experiments. Biometrika 39:419-22.

[12] ————. 1953. Accuracy of blood cell counting. Austral. J. Exp. Biol. Med. Sc. 31:603-6.

[13] ————. 1961. Significance tests in discrete distributions. J. Amer. Statist. Assoc. 56:223-34.

[14] ————. 1969. The Chi-squared Distribution. John Wiley & Sons, Inc., New York. Pp. xiv, 370.

[15] ————.; and Brown, T. A. I. 1965. Sizes of the χ^2 test in the symmetrical multinomials. Austral. J. Statist. 7:40-44.

[16] Moran, P. A. 1952. A characteristic property of the Poisson distribution. Proc. Camb. Philos. Soc. 48:206-7.

[17] Plum, P. 1936. Accuracy of haematological counting methods. Acta Med. Scandinav. 90:342-64.

[18] Sukhatme, P. V. 1937. The problem of k samples from a Poisson population. Proc. Nat. Inst. Sci. India 3:297-305.

[19] ————. 1938. On the distribution of χ^2 in samples of the Poisson series. J. Roy. Statist. Soc. 5:75-9.

[20] Stavem, P. 1964. The distribution of erythrocytes and reticulocytes in blood smears. Acta Med. Scandinav. 175:1-111.

[21] "Student" (W. S. Gosset) 1907. On the error of counting with a haemocytometer. Biometrika 5:351-60.

[22] Turner, M. E., and Eadie, G. S. 1957. The distribution of red block cells in the hemacytometer. Biometrics 13:485-95.

[23] Uppuluri, V. R. R. 1969. Likelihood ratio test and Pearson's χ^2 test in multinomial distributions. To be published in the Bulletin Internat. Statist. Instit.

[24] ————.; and Bowman, K. O. 1966. Likelihood Ratio Test Criterion for Small Samples from Multinomial Distributions. Oak Ridge Nat. Lab. Oak Ridge. Pp. vi, 47.

APPLICATION OF DISCRETE DISTRIBUTION THEORY TO THE STUDY OF NONCOMMUNICABLE EVENTS IN MEDICAL EPIDEMIOLOGY

P. FROGGATT
Department of Social and Preventive Medicine
The Queen's University
Belfast, Northern Ireland

SUMMARY

The abilities of five hypotheses ('chance,''proneness' or 'false contagion,' and three of 'true contagion') to describe data from three classes of noncommunicable events of importance in medical epidemiology (consultations in general practice, one-day industrial absences, and accidents) and covering more than one period of observation, are tested using maily discrete distribution methods. The hypotheses generate the Poisson, the negative binomial, and three other compound Poisson distributions, respectively; and these are fitted by maximum-likelihood methods. Discussion covers the interpretation, implications, and practical applications of the results. The arguments and methods are set out in a form intended to be readily understood by workers in the applied field and which should allow their application to contingent classes of discrete data.

1. INTRODUCTION

Medical epidemiology is increasingly concerned with 'non-communicable' (i.e., noninfectious), as compared to 'communicable' disease, and in the former class with events which may repeat (so that individuals incur 0,1,2,...) e.g., nonfatal accidents or episodes of sickness, as compared to events which do not, e.g., death. With the exception of accidents, however, such events have seldom been studied by examining inter alia the form of the frequency distributions which generate the conventional rates, despite the extra information which this may yield (Froggatt [19]).

In this paper we apply, to sets of our own data, curve-fitting and other techniques to test the adequacy of certain plausible hypotheses; we then examine how far the results contribute to our understanding of causation or, at the least, how far the data may be used empirically to predict, from his previous record, an individual's future experience with respect to the same event. The events we examine are of two general classes: first, those in which a considered decision by the individual is a crucial determinant of the outcome, viz., doctor/patient consultations in general practice and one-day absences from work; and second, those in which generally it is not, viz., traffic accidents incurred by public transport drivers. (For convenience these are termed throughout: 'consultations,' 'one-day absences,' and 'accidents'). For simplicity the fewest requisite number of (characteristic) sets of our extensive data for each event are considered here; details of the complete data and relevant literature have been fully described elsewhere by [21] for consultations, by [18] for one-day absences, and by [8] for accidents.

We adopt the following presentation. First we set out the hypotheses under test and name the theoretical models which their assumptions generate; next we present the observed and theoretical frequency distributions and their statistics and the results of some further relevant analyses; and finally we discuss the interpretation, implications, and practical application of the results. Throughout, we have endeavored to present the techniques and arguments in a form which should be readily understood by the epidemiologist, who is generally more concerned with the valid application and interpretation, rather than with the theoretical basis or properties of analytical methods.

2. HYPOTHESES AND THEORETICAL DISTRIBUTIONS

The hypotheses under test (A to E, below) and the theoretical models generated on their assumptions are now described. We do not give the statistical derivation of the models--these have been considered in the cited works; we do, however, emphasize that different assumptions may generate exactly similar models, and this leads to interpretative difficulties as discussed in Section 4, below.

Hypothesis A.

Random allocation of events in a homogeneous population in an environment either stable or which changes equally for all subjects.

Theoretical distribution: the Poisson ([45], [35], and [17]).

Hypothesis B.

Ab initio differences in the 'liability' of subjects, in an otherwise homogeneous group, to incur an event, the environment being either stable or changing equally for all subjects. 'Liability' (conventionally symbolized λ) follows a Pearson Type III distribution over the group, and each individual's λ remains unchanged. (Though the value of λ for a given individual is hypothesized as strictly constant over the (usually relatively short) period of study, with the passing of time some slight change is generally assumed; see, e.g., [4]. This is the classical hypothesis of 'unequal liabilities' ([24]); for convenience it will be given below its more usual synonym of 'proneness'.

Theoretical distribution: the negative binomial ([24], [42], and [35]).

Hypothesis C.

1. In an environment either stable or changing equally for all, every subject is equally liable to 'spells' (periods of time) within which all events he sustains must occur. The number of spells in a given time is a Poisson variable with parameter λ: this implies (a) that the number of spells incurred by an individual in two periods are independent; and (b) that the length of a spell is short in relation to the period of time under consideration.

2. The probability of an event occurring within a spell is constant and not dependent on the individual: thus accidents within spells are assumed to have a Poisson distribution with constant parameter θ.

3. No events can occur outside spells.

Theoretical distribution: the Neyman Type A ([43], [8], and [32]).

Hypothesis D.

Similar to Hypothesis C, though now relaxing assumption (3) to allow events to occur outside a spell, in which case they are independently distributed as a Poisson variable with parameter ϕ over the given time period: thus they occur independently both of one another and of accidents within spells.

Theoretical distribution: the 'Short; ([8] and [32]).

Hypothesis E.

Events occur in clusters which are randomly distributed among the subjects at equal risk. Only one or two events can occur in each cluster, and such events follow a binomial distribution within clusters. Clusters and events are independent. Randomly distributed events can occur which are not members of a cluster, but these cannot be identified from the distribution.

Theoretical distribution: the 'Hermite' (two-parameter form) ([33] and [31]).

A priori Hypotheses A ('chance') and B ('proneness') are coherent for each event studied here. So also are Hypotheses C and D if we can accept that 'spells' relate to, say, periods of minor ill-health--which directly produce one-day absences and consultations, or traffic accidents (if, as a result, the performance of the driver is substandard)--during which an individual incurs all his events (Hypothesis C) or some of them with the others as independent random phenomena (Hypothesis D). Hypothesis E is relevant particularly to consultations in that, although 'clusters' of consultations are themselves unlikely to follow random law, rarely more than two consultations (initial consultation and routine review) are necessary for the common minor complaints.

Many other hypotheses which are plausible in the context and which generate distributions with well-studied properties (e.g., those described in [4], [5], [11], and [49] could have been examined, but in practice some restriction on numbers had to be imposed.

3. TESTS OF THE HYPOTHESES

3.1 Evidence from Fitting the Distributions.

The theoretical distributions were fitted to each of the three sets of data (except, for technical reasons produced by the very long tail, the Short distribution, to consultations) using in all cases computer-derived maximum-likelihood (M-L) estimates of the distribution parameters and a chi-square goodness-of-fit test. The negative binomial and Neyman Type A were fitted from straightforward programs based on solutions given by [27] and [6] for the former and by [46] and [9] for the latter; for the Short and Hermite we used Algol programs compiled on technical information previously given by ([32] and [33])*. Computing was on the S.R.C. Atlas and the University I.C.L. 1905. In all instances satisfactory convergence of the parametric estimates was achieved within the programs' iterative instructions. The frequency distributions with means and variances are in Tables 1(a) - 1(c); the M-L estimates of the parameters with asymptotic standard errors are in Table 2.

For consultations (Table 1(a)), only the negative binomial is an acceptable model; for one-day absences (Table 1(b)) and accidents (Table 1(c)) all but the Poisson are adequate. (We will not further consider Hypothesis D because, despite its good fit, the Short distribution has (1) estimators with very large sampling errors which make their interpretation largely meaningless (and in terms of the physical model, the negative values of ϕ (Table 2) are, in fact, illegal), and (2) very high intercorrelations leading to difficulties in interpretation; e.g., for the data in Table 1(c) [32] gives values of $r_{\hat{\lambda}\hat{\theta}} = r_{\hat{\lambda}\hat{\phi}} = -0.997$, and $r_{\hat{\theta}\hat{\phi}} = +0.990$). As already emphasized, however, a successful fit with a model does not mean that the assumptions generating it necessarily explain the distribution of the events; to study this further and to select the most appropriate model from those successfully fitting the distributions (negative binomial, Neyman Type A, and Hermite) of one-day absences and accidents, additional evidence must be adduced.

3.2 Evidence from Correlation.

Further information is provided by the correlations between

---

* Programs were kindly made available by Mr. C. D. Kemp, who also facilitated this study in other ways. (See also pp. 41-65.)

the numbers of events the sample members incur in two equal non-overlapping periods of time. The value of the correlation coefficient should not differ from zero on Hypotheses A and C ([29])--and its value on Hypothesis D has not yet been established ([34]); but on Hypothesis B (proneness) it should be significantly different from zero and, on its strictest interpretation, independent of the periods selected, these results following from the postulated invariance of each individual's 'liability' (λ). (The nature of such correlational evidence in searching for 'proneness' has been discussed by, e.g., [40], [38], and [7], and the arguments reviewed by [17]). Since, however, the overt consequences of λ (consultations, one-day absences, accidents) are the units, we note that these results strictly follow only if the effect of any factor, not ascribable to λ and acting nonsystematically, is unimportant: "[On accident data] the interperiod correlation coefficients tend to fall as the interval between the two exposure periods increases--perhaps one might expect this on the 'proneness' hypothesis: any correlation due to the personal factor would tend to get more and more diluted by increasing changes in environmental conditions of a nonsystematic nature, affecting different subjects differently, and thus increasing the 'chance' component" (Irwin [29]).

Table 3 displays the relevant sample correlation coefficients (r) and their standard errors derived from normal theory. Each coefficient is positive, for each event their values are greater for contiguous than for noncontiguous periods, and those for consultations and one-day absences--which are mainly 'voluntary' events--greatly exceed those for accidents; and these general conclusions hold also when it has been possible to transform the generating distributions to ones approximating the normal distribution ([18]). These findings are consistent with proneness, but not with the other hypotheses studied. Proneness also requires reasonable stability of the interperiod coefficients. Testing [47], pp. 178-9) shows discordance of the sample coefficients in Table 3 for consultations ($\chi^2(2) = 39.02$, $P < 0.001$) but not for one-day absences ($\chi^2(2) = 4.61$, $0.20 > P > 0.10$) or accidents ($\chi^2(2) = 0.59$, $0.70 > P > 0.50$). Some differences in the interperiod coefficients are the rule and have been noted in the literature--for sickness episodes or absence from work by [48], [36], [50], and [18]; for industrial accidents, [13] and [15]; and for traffic accidents, [14], [26], and [8]: provided, however, that these differences are not marked and having regard to real-life conditions, proneness need not be rejected ([29]).

The analysis may be carried further by first deriving, from the parameters of the negative binomial distribution for events in a single period, the value which ρ--the correlation coefficient between the numbers of events incurred in that and any other similar period (in practice usually the one following)--would be expected to take if proneness completely explained the facts, i.e., postulating "proneness mixture--no contagion--no time-effect" in ([4] and [5]), terminology and then (1) comparing ρ with r, and (2) using ρ in regression to compare predicted with observed future experience. (These assumptions lead to the symmetrical bivariate negative binomial (S.B.N.B.) as the bivariate model with ρ also calculable from parameters k for the double-period and $m = \frac{1}{2}M$, where M is the double-period mean. This approach, however, is *a posteriori* and, moreover, uses data from the very period (the second) for which prediction is to be made. Establishing initially the adequacy of the S.B.N.B. for the first period's data (by dividing the period into two equal sub-periods) and from this inferring its success for the longer double period--as recommended by [2], [3] can be a worthwhile first step). This approach based on a single period's data has obvious value for prediction and has been studied extensively on the above and related assumptions (e.g., [42], [44], [37], [2 - 5], [39], and [49]; the treatment below is a simple application of the theory.

Following Newbold [42], Arbous and Sichel [2] reach the solution

$$\rho = \frac{m}{m + k},$$

where m and k are the univariate negative binomial parameters from the first period's data (they also derive the standard error of $\hat{\rho}$), and give the linear regression

$$\tilde{y} = \rho x + \rho k,$$

where $\tilde{y}$ is the predicted mean number of events in the second period for those with x = 0,1,2... in the first. Using M-L estimators $\hat{k}$ and $\hat{m}$ for the first period, we reach values of $\tilde{y}$ to compare with actual array means ($\bar{y}$), $\bar{y}$ following a linear regression in each set as required for the comparison (for establishment of linearity see: for consultations, [21], Table VIII; for one-day absences, [18], Table X.24(d); and for accidents, [8], Table 6.8)*. Table 4 shows that, for consultations and one-day absences, agreement between $\tilde{y}$ and $\bar{y}$ is good, but for accidents it is not, $\tilde{y}$ consistently overestimating $\bar{y}$. Thus for the first two classes of data, viz., events in which a considered decision is a determinant, but not for accidents, reasonable prediction of a second period's experience can be made from that of a previous (contiguous) one.

4. DISCUSSION AND CONCLUSIONS

4.1 Interpretation of the Results.

The results show that, of the hypotheses under test, Hypothesis B (proneness) explains the data best, and for consultations it is the only one tenable. Results of further testing, outside the scope of this paper to the present, also show that, while the S.B.N.B. is an acceptable model for one-day absences ([18]) and accidents ([8])--though not consultations ([21])--among most of the sample populations studied, it generally provides a less successful fit than the univariate form. In its strictest interpretation, therefore, proneness cannot be a complete explanation of the facts, especially of consultations. As is well known, however, the negative binomial may be generated on hypotheses other than proneness (see, e.g., [28] and [1]--"and if we had a negative binomial and it was a good fit [accident] proneness may be involved or it may not." (Greenwood [22])-- and consideration of the events under study suggests that at least the following four confounding phenomena may have been operating whether or not 'proneness' was present at all: (a) unequal 'exposure to risk' among the group members; (b) the probability of ascertaining an event not being independent of the person who incurs or records it; (c) a spells phenomenon (e.g., as in Hypotheses C, D, and E) operating in some instances; and (d) so-called 'contagion' in that incurring an event <u>ipso facto</u> makes that individual more (or less) likely to incur another.

As regards (a), under heterogeneous Poisson sampling, by

---

*On our complete data $\hat{\rho}$ generally underestimates r for one-day absences and accidents and overestimates it for consultations. Results (and explanations) are not given here; the regression results are considered more relevant.

which the negative binomial as the proneness model is derived, if the study group comprises several subgroups of different 'exposure to risk' and events are distributed among members of each subgroup (who are at true equal risk) as a Poisson variable with parameter different for each subgroup, the overall distribution could be fitted by the negative binomial without proneness necessarily operating at all. This holds for any heterogeneity in the group for a variable which influences the frequency of the event studied. Great care was taken in our source data to minimize this confounding effect; any influence due to it should be unimportant.

As regards (b), this operates classically as the 'tendency to report, (which confounds the 'tendency to have') phenomenon which has bedevilled much research into the personal characteristics of the 'accident-prone' based on--as is customary--records of minor industrial accidents. Again, on heterogeneous Poisson sampling this could lead to a distribution of events closely approximated by a negative binomial when, in fact, the underlying causes follow some different law. This should not be crucial to the data presented here, where the accidents are nearly completely--and the consultations and one-day absences completely--ascertained (though bias introduced by multi-doctor recording may confound the consultation data to some extent (see [21] for discussion)).

As regards (c), if spells have the meaning given in Hypothesis C, then under certain constricting conditions, viz., events within spells to be distributed as a logarithmic distribution but with added zeros, while the distribution of spells remains a Poisson, the overall distribution can be negative binomial in form (Kemp [32]). A successful fit with this model is therefore interpretable in terms either of 'spells' or 'proneness.' Also, Irwin [29] has noted that a successful fit with the Neyman Type A (generated here on a spells assumption)--as is the case with one-day absences and accidents in Table 1(b) and (c)--can also be given a proneness interpretation if we regard individuals as having 'pronenesses' $0, \theta, 2\theta, 3\theta, \ldots$, where θ has the meaning in assumption (2) to Hypothesis C. Such ambiguity in interpretation hardly allows us to distinguish between 'proneness' and 'spells' from curve-fitting (with these models) alone; and unfortunately it is a feature of contagious distributions that more than one set of basically different assumptions can lead to exactly the same model. Furthermore, we note that, in consultations in general practice, an unusual class in that episodes of illness and consultations within episodes can be severally identified from planned data, episodes of illness are not Poisson distributed [30] as assumed under Hypotheses C, D, and E, so that the overall distribution of consultations would be affected even if the distribution of consultations within episodes actually followed the assumptions in these hypotheses. In view of the extremely bad fit to the consultation data in this paper with the models generated on Hypotheses C and E (Table 1(a)) we are currently conducting such a planned prospective study which will allow us to decide which of the above statistics is more discordant with its formulated assumptions.

As regards (d), if all subjects have ab initio equal 'liability' (λ) to incur an event, but after a subject has incurred n events his 'liability' to have further ones changes so that λ per unit of time is a linear function of n, i.e., an hypothesis of 'contagion,' the resultant distribution can again be fitted by the negative binomial (see, e.g., [1], and more general treatment by [5]). Greenwood and Yule [24], when studying accidents, considered the general case of contagion--their 'biased' scheme which they derived in a general form but unsuitable for computation--but they did not notice that the negative binomial could be obtained from it ([29]). They simplified the scheme, as a working model, to their 'modified biased' distribution which allowed only one change in liability (a decrease--after the first accident) irrespective of the type of accident and no matter how many subsequent accidents were incurred. Others (among them [37], [28], [16], [35], and [5]) also considered the argument and arrived at the distribution law in the form of a negative binomial with the same parametric form to the proneness model. Thus 'contagion' and 'proneness' cannot be differentiated solely from a univariate negative binomial fit: theoretically they may be distinguished from the corresponding bivariate models, but these have only slightly different parametric forms, and in practice only an extremely sensitive test could discriminate between them [35], [5], and [17]. We note, however, that on this contagious hypothesis the distribution mean would increase (or decrease) over successive periods if every individual's liability increased (or decreased) under the stated assumptions; or, if the liability of some increased and others decreased, the mean, though not the variance, could remain unchanged. These are controverted by the present data (including scrutiny of the individuals' records) over observation periods of three years for consultations, up to seven years for one-day absences, and four years for accidents. Thus contagion ex hypothesi is unlikely to be operating generally to an important extent and seems a far too unsophisticated model for any of the three classes of data considered. Contagion may, however, in a more general sense, operate on some individuals and act to dilute the proneness component. This would contribute to the generally less successful fit of the S.B.N.B. compared to the univariate model, as already noted, and thus to the disparities between the values of r and $\hat{\rho}$--especially as the two observation periods become more remote from each other--as demonstrated on the extensive data in the source references. (Where a more sensitive test is required, discordance between the observed cell frequencies on the bivariate table with those expected on the S.B.N.B.--as calculated from the joint probability density function--can be tested by chi-square). Nevertheless, as a general theory and despite its rigid assumptions, proneness is markedly successful in explaining all three classes of data examined here.

4.2 Implications of the Findings.

If proneness be accepted we should consider its coherence and implications. In the field of accidents, from which the hypothesis is borrowed, λ (liability) was assumed to be interpretable physically as the (stable) net result of personal characteristics (christened 'proneness' by Farmer and Chambers [12] and called 'personal tendency' by Newbold [42]) and environmental factors: thus λ "contains, in most practical cases, as well as the actual personal tendency, any constant external differential risk affecting particular persons which may exist in the same occupational group.... We cannot, of course, by any statistical modification distinguish between that part of the λ which is personal and that which depends upon constant external bias" ([42]). By studying groups of subjects homogeneous (as far as possible) for environmental factors, it was argued that λ would measure proneness; and further, since λ and its overt consequence (accidents in the context) would necessarily be related, constant differences between subjects in the latter would be ascribable to differences in their proneness after allowance for extraneous phenomena.

Identification of individuals with high proneness--inferred *ex post facto* from their accident record--and the continuing, though generally unsuccessful, search for social and biological correlates, naturally follows from this reasoning. (Detailed treatment of the 'accident-proneness' concept and research is given by Froggatt and Smiley [20]; see also the article by C. D. Kemp in this volume.)

With consultations and one-day absences, however, which have many distinguishable causes, it is unrealistic to conceive *a priori* such a narrow underlying common factor as proneness--certainly in its strict classical sense of "a special personal susceptibility inherent in the individual and differing from one individual to another" (Newbold [41], Preface). It is also unnecessary to do so. As generated here, λ, in the negative binomial, can simply represent the net effect of many factors contributing to, though not necessarily exclusively causing, the event in question: for each subject these factors need not each be a constant though their net result must *ex hypothesi*. λ, therefore, has its original, wider, historical connotation as comprising "a motley host of motives and factors which will be very difficult indeed to separate and measure" ([23]). The components of this 'motley host' cannot readily be identified nor even categorized coarsely as 'personal' or 'environmental': for example, a consultation typically involves contracting an ailment, deciding to consult or not consult the doctor, and obeying or not his instructions for any subsequent attendance; and a one-day absence is also determined by an equally complex chain of events. Moreover, important contributing factors must exist which are neither true random phenomena nor strictly components of λ; since these cannot be identified in the real-life situation we cannot decide what links in the complex chain of causation should be completely ascribable to λ. These are formidable problems in physical interpretation and it seems unlikely that further elucidation is possible from the data alone.

Irrespective of its components we may, however, determine how precisely λ can be measured from the individual consultation, one-day absence, or accident record. Accepting the S.B.N.B. as the model and the assumptions here generating it, we may calculate the posterior distribution of λ for given x (the number of events in the double-period). Amending Kerrich [35], eq. 5.23, to our symbolization, we obtain for the data:

$$\nu = 2(\frac{k}{m} + 2)\lambda,$$

which has a χ^2 distribution with 2(k + x) degrees of freedom. Evaluating with $k = \hat{k}$ for the double-period, and $m = \frac{1}{2}\hat{M}$ (where $\hat{M}$ is the mean of the double-period) we reach estimates for the limits of λ (at P = 0.95 and 0.05 on the table of χ^2) when x = 0,1,2,... (Table 5). Following [35], eq. 5.16, we estimate $\rho_{\lambda x}$ for consultations, one-day absences, and accidents as, respectively, 0.928, 0.798 and 0.581, evidencing concomitant variation between λ and x; but the 90 per cent limits of λ, for given values of x, considerably overlap each other (Table 5) so that, for example, we could not assert that an individual having (say) four consultations over the period had necessarily a higher 'liability' to incur a consultation than a person having none. (As already noted the S.B.N.B. is strictly not always an acceptable model, particularly for consultations; but the (univariate) M-L estimators in Table 2 and the (bivariate) ones footnoted in Table 5 show such agreement that this would have little effect on the limits of λ and none at all on the argument). Therefore, we have a factor, λ, whose components we cannot identify and quantitate from the data and which is itself in some ways imprecisely measured on the criterion of the number of events observed. This is the stark situation facing researchers into 'proneness' in every human field of enquiry.

With consultations and one-day absences--in which a considered decision is the ultimate determinant--we note that λ itself, whatever may be its components, correlates better with the actual events than is the case with accidents. It may be speculated that with these two events this is due to a major component of λ being the relatively stable (over the period) attitude, make-up, or circumstances of the individual which, either alone or in interaction with factors of the *milieu extérieur*, *largely determines the decision itself*: with consultations, perhaps a 'neurotic' temperament, overconcern with one's health, or poor health *per se*; with one-day absences, perhaps also a certain lack of corporate loyalty or conscientiousness. 'Personal' factors of this type would not generally determine accidents though they may (and often do) correlate weakly with them, but they would *a fortiori* largely determine those classes of mainly 'voluntary' events to which consultations and one-day absences respectively belong.

4.3 Concluding Remarks.

It seems unlikely that discrimination between acceptable hypotheses can be made from the data alone since no successful model so far developed is unequivocal in interpretation. Moreover, complex theoretical distributions, derived on more refined and flexible hypotheses, will generally be compound Poissons (in Gurland's [25] terminology) with three or more parameters to be estimated from the data: since these estimates will probably be unstable, the value of such models will be severely limited if one intends to attach some physical meaning to the parameters, as would generally be the case. In fact it could be held that no mathematical model could reasonably explain such a complex situation as one involving human reactions. Identification of individuals with 'high' and 'low liability'--as inferred ex post facto from their record of the events studied--followed by their examination and interview to search for primary group differences in some social and biological parameters, would seem a logical line of further research. This has been energetically pursued in accidents (see [8] for review) but only occasionally in industrial absence (e.g., [50]) and consultations ([10]).

Irrespective of their 'cause' it would be an undeniable advantage to an employer (for one-day absences and accidents) or a doctor (for consultations) to be able to predict the likely future experience of an individual with respect to the event in question--ideally from some measureable personal quality identifiable at the outset. This has not so far been possible: the associations between personal characteristics and the numbers of events incurred have been too weak. Pragmatic business and practice managers can note, however, that as evidenced in Table 4 reasonable prediction of future experience can be made from that of the past in the case of one-day absences and consultations, though perhaps not to a power, on such short observational periods, which would justify executive action through a tolerably low misclassification rate. Nevertheless, for these two events the interperiod correlations (Table 3) are sizeable, larger, in fact, than is generally obtained between a personal characteristic and its overt consequence in any contingent field of human experience. Identifying components of λ, therefore, may be academic rather than of practical value. With accidents, on the other hand, the inter-period correlations are weak, which alone would justify searching, in practice, among personal characteristics of the individual for some stronger correlate. The value of $\rho_{\lambda x}$ (0.581, as above, which is representative of its usual order) suggests that, on the hypothesis, many accidents occur independently of the underlying proneness; and this, in turn, would indicate that any correlate identified will have severely limited value or application in practice. This, in fact, has been the case.

Despite interpretative and other difficulties, discrete distribution theory provides epidemiology with an important technique for the examination of repeated events. At the very least it allows hypotheses to be rejected and, in particular cases, can supply information of value in prediction; at the most it can identify a uniquely successful model (of those tested) for the distribution law, and, if the data be rigorous, complementary analyses can often suggest which known interpretation of the model is the more likely general explanation of the data in the particular circumstances. The techniques have been widely applied in contingent fields, for example, in biology and actuarial practice, and in medicine itself classically in the study of epidemics of infectious disease. With the exception of accidents, however, they have been largely (though not completely) ignored in the study of noncommunicable disease; and it is hoped that this paper, as well as outlining simple methods of application and general lines of interpretation, has indicated the value of discrete distribution methods to this last class of data.

ACKNOWLEDGEMENTS

I am indebted to Mr. C. D. Kemp, Department of Statistics, The Queen's University, Belfast, who made available his programs for fitting the Neyman Type A, Short, and Hermite distributions and who also arranged time on the Science Research Council Atlas computer at Chilton; and to his colleague, Mr. G. W. Cran, who facilitated some of the other computer operations.

REFERENCES

[1] Anscombe, F. J. 1950. Sampling theory of the negative binomial and logarithmic series distributions. Biometrika 37:358-82.

[2] Arbous, A. G., and Sichel, H. S. 1954a. New techniques for the analysis of absenteeism data. Biometrika 41:77-90.

[3] ————., ————. 1954b. The use of estimates of absence-proneness for guiding executive action. Appl. Statist. 3:159-73.

[4] Bates, Grace E., and Neyman, J. 1952a. Contributions to the theory of accident proneness. I: An optimistic model of the correlation between light and severe accidents. Univ. Calif. Publs. Statist. 1:215-54.

[5] ————., and ————. 1952b. Contributions to the theory of accident proneness. II: True or false contagion. Univ. Calif. Publs. Statist. 1:255:76.

[6] Bliss, C. I., and Fisher, R. A. 1953. Fitting the negative binomial distribution to biological data. Biometrics 9:176-200.

[7] Blum, M. L., and Mintz, A. 1951. Correlation versus curve fitting in research on accident proneness: Reply to Maritz. Psychol. Bull. 48:413-8.

[8] Cresswell, W. L., and Froggatt, P. 1963. The Causation of Bus Driver Accidents: An Epidemiological Study. London, Oxford University Press.

[9] Douglas, J. B. 1955. Fitting the Neyman type A (two parameter) contagious distribution. Biometrics 11:149-73.

[10] Dudgeon, M. Yvonne; Froggatt, P.; and Turkington, E. E. 1970. Consultations in general practice: Characteristics of 'frequent consulters'. In press.

[11] Edwards, Carol B., and Gurland, J. 1961. A class of distributions applicable to accidents. J. Amer. Statist. Ass. 56:503-17.

[12] Farmer, E., and Chambers, E. G. 1926. A Psychological Study of Individual Differences in Accident Rates. Med. Res. Coun., Industr. Fat. Res. Bd., Report No. 38. London, H.M. Stationery Office.

[13] ————, ————. 1929. A Study of Personal Qualities in Accident Proneness and Proficiency. Med. Res. Coun., Industr. Hlth. Res. Bd., Report No. 55, London, H.M. Stationery Office.

[14] ————, ————. 1939. A Study of Accident Proneness amongst Motor Drivers. Med. Res. Coun., Industr. Hlth. Res. Bd., Report No. 84. London, H. M. Stationery Office.

[16] Feller, W. 1943. On a general class of 'contagious' distributions. Ann. Math. Statist. 14:389-99.

[17] Fitzpatrick, R. 1958. The detection of individual differences in accident susceptibility. Biometrics 14:50-68.

[18] Froggatt, P. 1967. Short-term Absence from Industry: A Statistical and Historical Study. 2 vols. Thesis for degree of Doctor of Philosophy. Belfast, The Queen's University.

[19] ———. 1968. Research in industrial morbidity: principles and opportunities. Trans. Soc. Occup. Med. 18:89-95.

[20] ———, and Smiley, J. A. 1964. The concept of accident proneness: a review. Brit. J. Industr. Med. 21:1-12. (See also C. D. Kemp, this volume.)

[21] ———; Dudgeon, M. Yvonne; and Merrett, J. D. 1969. Consultations in general practice: analysis of individual frequencies. Brit. J. Prev. Soc. Med. 23:1-11.

[22] Greenwood, M. 1949. Discussion on Dr. Smeed's paper (Some statistical aspects of road safety research). J. R. Statist. Soc. A 112:25.

[23] ———, and Woods, Hilda M. 1919. A Report on the Incidence of Industrial Accidents upon Individuals with Special Reference to Multiple Accidents. Med. Res. Coun., Industr. Fat. Res. Bd., Report No. 4. London, H.M. Stationery Office.

[24] ———, and Yule, G. U. 1920. An inquiry into the nature of frequency distributions representative of multiple happenings, with particular reference to the occurrence of multiple attacks of disease or repeated accidents. J. R. Statist. Soc. 83:255-79.

[25] Gurland, J. 1957. Some interrelations among compound and generalized distributions. Biometrika 44:265-8.

[26] Häkkinen, S. 1958. Traffic Accidents and Driver Characteristics: A Statistical and Psychological Study. Institute of Technology Scientific Researches, No. 13. Helsinki, Finland's Institute of Technology.

[27] Haldane, J. B. S. 1941. The fitting of binomial distributions. Ann. Eugen., Lond. 11:179-81.

[28] Irwin, J. O. 1941. Discussion on Chambers and Yule's paper (Theory and observation in the investigation of accident causation). J. R. Statist. Soc. Suppl. 7:101-7.

[29] ———. 1964. The personal factor in accidents: a review article. J. R. Statist. Soc. A 127:438-51.

[30] Jacob, A., and Pearson, J. 1967. Morbidity in an 'artificial practice'. J. Coll. Gen. Practit. 13:303-12.

[31] Kemp, Adrienne W., and Kemp, C. D. 1966. An alternative derivation of the Hermite distribution. Biometrika 53:627-28.

[32] Kemp, C. D. 1967. On a contagious distribution suggested for accident data. Biometrics 23:241-55.

[33] Kemp, C. D., and Kemp, Adrienne W. 1965. Some properties of the 'Hermite' distribution. Biometrika 52:381-94.

[34] ————, ————. 1969. Personal communication.

[35] Kerrich, J. E. 1951. Accident statistics and the concept of accident proneness. Part II: The mathematical background. Biometrics 7:391-432.

[36] Lokander, S. 1962. Sick absence in a Swedish company: a socio-medical study. Acta. med. Scand. Suppl. 377:1-172.

[37] Lundberg, O. 1940. On Random Processes and their Application to Sickness and Accident Statistics. Uppsala, Almquist & Wiksells.

[38] Maritz, J. S. 1950. On the validity of inferences drawn from the fitting of Poisson and negative binomial distributions to observed accident data. Psychol. Bull. 47:434-43.

[39] Mellinger, G. D.; Sylwester, D. L; Gaffey, W. R; and Manheimer, D. I. 1965. A mathematical model with applications to a study of accident repeatedness among children. J. Amer. Statist. Assn. 60:1046-59.

[40] Mintz., A., and Blum, M. L. 1949. A re-examination of the accident proneness concept. J. Appl. Psychol. 33:195-211.

[41] Newbold, Ethel M. 1926. A Contribution to the Study of the Human Factor in the Causation of Accidents. Med. Res. Coun., Industr. Fat. Res. Bd., Report No. 34. London, H.M. Stationery Office.

[42] ————. 1927. Practical applications of the statistics of repeated events, particularly to industrial accidents. J. R. Statist. Soc. 90:487-535.

[43] Neyman, J. 1939. On a new class of 'contagious' distributions applicable to entomology and bacteriology. Ann. Math. Statist. 10:35-57.

[44] Pólya, G. 1930. Sur quelques points de las théorie des probabilités. Ann. de l'Institut Henri Poincaré 1:117-61.

[45] Poisson, S. D. 1837. Recherches sur la Probabilité des Jugements. Paris, Bachelier.

[46] Shenton, L. R. 1949. On the efficiency of the method of moments and Neyman's Type A distribution. Biometrika 36:450-4.

[47] Snedecor, G. W. 1956. Statistical Methods. 5th edition. Ames, The Iowa State College Press.

[48] Snow, E. C. 1913. Some statistical problems suggested by the sickness and mortality data of certain of the large Friendly Societies. J. R. Statist. Soc. 76:445-517.

[49] Subrahmaniam, K. 1966. A test for 'intrinsic correlation' in the theory of accident proneness. J. R. Statist. Soc. B 28:180-9.

[50] Taylor, P. J. 1968. Personal factors associated with sickness absence. Brit. J. industr. Med. 25:106-18.

Tables 1a - 1c. Observed Frequencies (f_x) and Those Expected on the Poisson (P), Negative Binomial (N.B.), Neyman Type A (N.T.A.), Short, and Hermite Distributions using Maximum - Likelihood Parametric Estimates.

Table 1a. Surgery Consultations of 2810 Female Patients Registered in One Group Practice throughout 1962.
Mean = 2.768 Variance = 11.805

x	f_x	P	N.B.	N.T.A.	Hermite
0	820	176.44	818.99	813.08	503.83
1	535	488.38	533.78	337.53	337.28
2	369	675.90	378.00	390.25	641.55
3	283	623.63	274.78	345.12	379.09
4	201	431.55	202.33	271.76	400.02
5	149	238.90	150.13	203.59	212.66
6	106	110.21	111.96	147.60	163.64
7	76	43.58	83.79	103.68	79.40
8	77	15.08	62.88	70.67	49.57
9	54	4.64	47.29	46.90	22.20
10	32	1.28	35.62	30.41	11.89
11	31	0.32	26.87	19.31	4.96
12	27	0.08	20.29	12.04	2.36
13	14	0.02	15.33	7.37	0.92
14	3		11.60	4.44	0.40
15	6		8.78	2.64	0.15
16	8		6.65	1.55	0.06
17	3		5.04	0.89	0.02
18	2		3.82	0.51	0.01
> 19	14		12.05	0.64	
Total	2810	2810.01	2809.98	2809.98	2810.01
χ^2(D.F.)*		> 1000(8)	17.27(16)	328.95(13)	> 1000(10)
P		< 0.001	0.30-0.50	< 0.001	< 0.001

*For the calculation of χ^2 in this table and Tables 1(b) and 1(c), expected frequencies are pooled to give a minimum value of 4.0. The number of degrees of freedom = n - (number of parameters +1).

Table 1b. One-day Industrial Absences (All Causes) during 1957 of 195 Non-supervisory Married Male Inspectors Employed In One Company Throughout the Year.

Mean = 2.795 Variance = 4.608

x	f_x	P	N.B.	N.T.A.	Short	Hermite
0	27	11.92	23.38	25.78	26.26	25.11
1	35	33.31	38.52	36.32	34.65	32.76
2	35	46.55	39.64	38.03	38.39	40.08
3	37	43.37	32.62	32.38	32.76	33.71
4	19	30.30	23.48	24.07	24.58	25.92
5	20	16.94	15.45	16.16	16.44	16.81
6	10	7.89	9.53	10.01	10.09	10.09
7	6	3.15	5.59	5.81	5.76	5.46
8	3	1.10	3.16	3.19	3.09	2.77
9	3	0.34	1.73	1.67	1.57	1.31
> 10	0	0.13	1.91	1.58	1.41	0.97
Total	195	195.00	195.01	195.00	195.00	194.99
χ^2(D.F.)		39.57(6)	4.36(6)	3.02(6)	2.93(5)	3.92(6)
P		< 0.01	0.5-0.7	0.8-0.9	0.7-0.8	0.5-0.7

Table 1c. Traffic Accidents Incurred During 1952-1955 by 708 Public Corporation Bus Drivers Driving Regularly Throughout the Period.

Mean = 2.292 Variance = 3.446

x	f_x	P	N.B.	N.T.A.	Short	Hermite
0	117	71.53	111.13	115.97	116.88	118.73
1	157	163.97	168.60	162.25	160.43	151.83
2	158	187.94	156.41	153.56	153.64	157.25
3	115	143.61	114.36	115.57	116.05	118.33
4	78	82.30	72.38	74.67	75.13	77.68
5	44	37.73	41.54	43.09	43.29	43.85
6	21	14.42	22.21	22.75	22.76	22.47
7	7	4.72	11.25	11.17	11.09	10.46
8	6	1.35	5.46	5.16	5.07	4.52
9	1	0.35	2.56	2.26	2.20	1.82
10	3	0.08	1.17	0.94	0.91	0.69
11	1	0.02	0.52	0.38	0.36	0.25
> 12	0		0.40	0.23	0.19	0.12
Total	708	708.02	707.99	708.00	708.00	708.00
χ^2(D.F.)		64.15(6)	3.46(7)	2.63(6)	2.56(5)	3.40(6)
P		< 0.001	0.8-0.9	0.8-0.9	0.7-0.8	0.7-0.8

Table 2. Maximum-Likelihood Parametric Estimates ($\hat{E}$) and their Asymptotic Standard Errors (σ) for the Theoretical Distributions in Table 1

Parameter		Table 1(a) $\hat{E}$	Table 1(a) σ	Table 1(b) $\hat{E}$	Table 1(b) σ	Table 1(c) $\hat{E}$	Table 1(c) σ
P	(m)	2.768	-	2.795	-	2.292	-
N.B.	m	2.768	0.065	2.795	0.110	2.292	0.070
	k	0.853	0.034	4.014	0.772	4.487	0.762
N.T.A.	λ	1.459	0.044	4.080	0.983	4.642	0.716
	θ	1.897	0.054	0.685	0.169	0.494	0.077
Short	λ	-	-	9.953	*	6.736	13.8
	θ	-	-	0.427	*	0.408	0.437
	ϕ	-	-	-1.453	4.406	-0.454	2.70
Hermite	a_1	0.670	0.034	1.305	0.299	1.279	0.137
	a_2	1.049	0.024	0.745	0.158	0.507	0.070

*not computed

Table 3. Correlation Coefficients (and their Standard Errors) for the Number of Events Incurred by the Same Individuals in Two Equal Non-Overlapping Periods of Time.

	Consultations		One-day Absences*		Accidents*	
Period	A	B	A	B	A	B
B	0.647 (0.014)		0.612 (0.053)		0.239 (0.048)	
C	0.551 (0.016)	0.642 (0.014)	0.488 (0.064)	0.585 (0.056)	0.198 (0.049	0.246 (0.048)

Consultations	: A = 1962	B = 1963	C = 1964
One-day absences	: A = 1956	B = 1957	C = 1958
Accidents	: A = 1952-3	B = 1954-5	C = 1956-7

* Based on smaller populations (142 and 380 persons, respectively) than in Table 1 because of the longer period for the data

Table 4. Actual ($\bar{y}$) and Predicted ($\tilde{y}$) Mean Numbers of Events per Person in the Second Period for Those Having x = 0,1,2... Events in the First.

x	Consultations		One-day Absences		Accidents	
	$\bar{y}$	$\tilde{y}^{*}$	$\bar{y}$	$\tilde{y}^{*}$	$\bar{y}$	$\tilde{y}$
0	0.98	0.65	1.52	1.65	0.77	1.29
1	1.66	1.42	1.88	2.06	0.95	1.48
2	2.19	2.18	2.20	2.47	1.07	1.67
3	2.87	2.95	2.46	2.88	1.25	1.86
4	3.54	3.71	3.52	3.29	2.00	2.05
5	4.26	4.48	4.65	3.70		
6	4.95	5.24	4.70	4.12		
7	6.25	6.01				
8	6.77	6.77				
9	7.06	7.54				

Periods of the data

Consultations	:	1962 and 1963	(2810 persons)
One-day absences	:	1957 and 1958	(195 persons)
Accidents	:	1952-3 and 1954-5	(708 persons)

* From parametric estimates in Table 2. Those for accidents are - $\hat{m} = 1.291$, $\hat{k} = 5.415$

Table 5. 90 Per Cent Limits for λ on the S.B.N.B. Distribution given that an Individual incurs x Events over the Double-Period of the Data.

x	λ*		
	Consultations (1962 & 1963)	One-day Absences (1957 & 1958)	Accidents (1952-3 & 1954-5)
0	<0.050-1.191	0.303-2.123	0.281-1.430
1	0.139-1.986	0.482-2.582	0.387-1.663
2	0.333-2.666	0.678-3.024	0.498-1.890
3	0.569-3.306	0.885-3.454	0.614-2.113
4	0.828-3.911	1.101-3.876	0.733-2.332
5	1.105-4.498	1.324-4.290	0.855-2.548
6	1.400-5.081	1.553-4.698	0.980-2.762
7	1.697-5.648	1.787-5.102	1.107-2.973
8	2.006-6.207	2.025-5.501	1.235-3.183
9	2.322-6.759	2.266-5.897	1.365-3.391
10	2.644-7.331	2.510-6.290	1.497-3.597

*M.L. Estimators

Consultations: $\hat{k} = 0.905$, $\frac{1}{2}M = 2.296$ (n = 2810)

One-day absences: $\hat{k} = 3.230$, $\frac{1}{2}M = 2.834$ (n = 195)

Accidents: $\hat{k} = 4.487$, $\frac{1}{2}M = 1.146$ (n = 708)

"ACCIDENT PRONENESS" AND DISCRETE DISTRIBUTION THEORY

C. D. KEMP
Department of Statistics
The Queen's University of Belfast
Northern Ireland

SUMMARY

In this paper the author propounds the view that the idea of accident proneness, while proving ineffectual as an operational concept, has stimulated much interesting statistical theory and method and that actuarial risk theory has developed independently along similar lines. A selected bibliography of nearly 150 entries is provided.

1. INTRODUCTION

During the present century much work has been devoted to the analysis of accident data--particularly to counts of numbers of accidents sustained by individuals over a given time period. A constantly recurring concept has been that of accident proneness.

The author's work on discrete distribution theory has led him to study the literature on accident proneness and related topics. From this he has drawn the following conclusions:

(1) from a practical point of view (e.g., in terms of its contribution to accident prevention), the concept of accident proneness has proved singularly ineffectual;

(2) from the standpoint of statistical theory, however, it has been very fruitful in that it has given rise to much work, both intrinsically interesting and of value in other fields; and

(3) accident proneness theory and actuarial risk theory have developed simultaneously along paralled lines with surprisingly few crosscontacts.

Within the limits imposed by the space available, this paper attempts to document this interpretation with especial reference to the contagious discrete distributions.

The bibliography consists almost entirely of papers to which the author has had access. The very few which have not been examined are marked by an asterisk and are usually specifically related to one of the other references. While this bibliography contains, it is hoped, all the major references to accident proneness which have any statistical implications, almost certainly the actuarial literature is underrepresented; and the reader's attention is drawn to Philipson [143] and [144], the second part of which contains a comprehensive bibliography. Arbitrary boundaries have also had to be imposed on the coverage of discrete distribution theory, and hence important areas (e.g., infinite divisibility) are not covered at all in this paper.

2. NOTATION AND DEFINITIONS

Use will be made of the probability generating function (p.g.f.)

$$g_X(s) = E(s^X) = \sum_x p_x s^x,$$

e.g., the basic Poisson distribution has $g(s) = \exp\{\lambda(s-1)\}$ and for the Poisson time process the p.g.f. is $\exp\{\lambda t(s-1)\}$.

Gurland's [81] definitions and notation will be adopted concerning compound and generalized distributions:

Compounding: Given real variables X_1, X_2, X_3, then $X_3 \sim X_1 \wedge X_2$ will imply

$$g_3(s) = \int_{-\infty}^{\infty} g_1(s \mid cx_2)\, dF(x_2),$$

where c is a scaling constant, e.g., t (time).

If X_1 has a Poisson distribution, then

$$g_3(s) = \int_{-\infty}^{\infty} e^{cx_2(s-1)}\, dF(x_2),$$

and X_3 has a compound Poisson distribution. X_2 has the compounding distribution (in risk theory, the unconditional risk distribution).

Generalizing: The notation $X_3 \sim X_1 \vee X_2$ implies that $g_3(s) = g_1[g_2(s)]$. If X_1 has a Poisson distribution, then X_3 has a generalized Poisson distribution with p.g.f.

$$g_3(s) = \exp[\lambda\{g_2(s) - 1\}].$$

Distributions with p.g.f.s of this form may be regarded from two differing viewpoints. First, suppose that X_1 represents the number of clusters (say) and has a Poisson distribution, and that X_2 represents the number of individuals in a single cluster. The X_3 will represent the total number of individuals.

Alternatively, let $g_2(s) = \sum_i \pi_i s^i$; then

$$g_3(s) = \exp[\lambda\{\sum_i \pi_i s^i - 1\}]$$

$$= \exp[\sum_i a_i(s^i - 1)], \text{ where } a_i = \lambda\pi_i,$$

and X_3 may be regarded as arising from the convolution of Poisson singlet (parameter a_1), Poisson doublet (parameter a_2), triplet, etc., distributions. Such a distribution is sometimes called a multiple Poisson distribution.

An important relationship between compound and generalized distributions is given by Gurland's theorem, which states that if X_1 has a p.g.f. of the form $\{h(s)\}^\lambda$, where λ is the random variable X_2, then

$$X_1 \wedge X_2 \sim X_2 \vee X_1.$$

In Haight's [135] terminology, a compound Poisson is a mixed Poisson distribution, and a generalized Poisson arises from Poisson mixing.

Gurland's terminology may be extended so as to relate to continuous as well as discrete variables.

3. 1900 - 1920

The idea of an accident (e.g., road accident) as a purely fortuitous event is still widely held. This idea closely underlay the pioneering work of Bortkiewicz [1] who used it to derive the binomial Poisson limit (with credit to Poisson) which he applied to data, _inter alia_, on fatal accidents at work, as well as deaths from horse kicks.

The concept of complete randomness also underlies the stochastic derivation of the Poisson distribution from the Poisson process postulates; this derivation appears in Bortkiewicz's last publication [8] on the Poisson distribution, in which an interesting application is made to deaths of members and honorary members of the International Statistical Institute. (This derivation had appeared earlier in telephone traffic studies, Erlang [2], and in connection with Geiger counts, Bateman [4], counts of blood corpuscles and counts of gas molecules, McKendrick [6]).

Timerding [9] also used the Poisson distribution as the basis for his study of accident causation.

Two of the earliest papers contradicting the concept of pure chance were those by Greenwood and Woods [12] and Greenwood and Yule [11] on industrial accidents during the First World War.

Greenwood and Woods [12] considered three hypotheses concerning such data:

(a) PURE CHANCE, which leads, as already seen, to the Poisson distribution;

(b) TRUE CONTAGION, i.e., the hypothesis that all individuals start with an equal chance of having an accident, but that this chance is modified by each accident sustained - the resultant distribution they called the 'biased distribution' (if the chance of an accident remains unaltered after the occurrence of the first accident, then the distribution is called the "burnt fingers' distribution);

(c) APPARENT CONTAGION, i.e., the hypothesis that individuals have constant but unequal chances of having an accident.

Greenwood and Yule first supposed that these varying chances could be considered as having a normal distribution, and they obtained the first few probabilities of the resultant Poisson $\wedge$ normal distribution. Kemp and Kemp [131] demonstrated the identity of this with the Hermite distribution. Here compounding is purely formal since the $(-\infty,\infty)$ range of the normal variable implies Poisson distributions with negative parameters. Greenwood and Yule wanted a physically meaningful model, and this implies a non-negative compounding variable. It is important to emphasize that they then chose the gamma (Pearson Type III) as the compounding distribution purely for mathematical convenience (a simple, tractible distribution on $(0,\infty)$). This, of course, yielded the negative binomial distribution.

This particular model for the negative binomial has been used subsequently in many different fields (see, e.g., Bliss and Fisher [71]). (A summary of the properties and applications of the distribution was given by Bartko [96].) However, certain other papers published during this period, e.g., Whitaker [7], advocating the use of the negative binomial in opposition to the Poisson distribution, did so on empirical, rather than rational, grounds.

Greenwood and Yule's derivations of the Poisson $\wedge$ normal and negative binomial distributions seem to be the first instances of compound Poisson distributions in the literature.

Greenwood and Woods [10] considered that general policy over accident prevention would vary greatly according to which hypothesis seems most likely. If (a) is true then research should be directed towards improving general conditions, but not into the psychology or physiology of individuals; if (c) is true care should be taken in the selection of individuals, and the removal of those sustaining several accidents should reduce the casualty rate.

However, the negative binomial distribution arises from a number of different models (several of which may be viewed as true contagion models): inverse sampling, Yule [5]; a stochastic birth and immigration process with zero initial population,

McKendrick [6]; a limiting form of the Polya distribution, Eggenburger and Polya [11, 19]; a simple birth process with nonzero initial population, Yule [13], Furry [34]; and also from other stochastic processes and as a generalized Poisson, as we shall see later. So a good fit of the negative binomial distribution to accident data is not convincing evidence of Greenwood and Yule's hypothesis (c) (unequal initial susceptibilities), and hence removal of individuals suspected of having a high susceptibility may not lower the accident rate. Note that, as Greenwood [54] himself states, such susceptibility (liability) had not yet, in 1920, been christened "accident proneness."

During this period Filip Lundberg in Scandinavia had started to create the collective theory of risk. Almost all of his early papers were written in Swedish, with the exception of Lundberg [3], which is in German, and were largely inaccessible to the English-speaking world. Cramer [14] and Laurin [22] gave reviews in English of Lundberg's work.

Lundberg replaced the older notion of individual risk theory (in which the total risk business of an insurance company is the sum of a large number of individual risk situations) by the idea of collective risk, obtaining the total number of claims received by the company as a stochastic (Poisson) process. He also seems to have been the first person to examine distributions convoluted together in such a way that the number of convolutions has a Poisson distribution; the resultant distributions are generalized Poisson distributions. A detailed explanation is given by Cramer [77], section 3.2.

Lundberg's classical form of the risk process is, by definition, independent of time, i.e., the risk distribution is time homogeneous. Although this is a somewhat unrealistic assumption, the simplicity of this approximation to reality led to a remarkable advancement in the state of risk theory.

4. 1921 - 1939

Further statistical investigations under the Industrial Fatigue Research Board into the study of accident prevention were carried on by Newbold [17, 18]. Both Newbold's [18] paper and the discussion on it are still well worth reading: she not only summarized the statistical results to date, but also studied the

the relationship between the susceptibility and resultant compound Poisson distributions, and carefully stated the limitations of the theory, as well.

Newbold worked in conjunction with the psychologists Farmer and Chambers, who first introduced the term "accident proneness" in Farmer and Chambers [15]: "the fact that one of the factors connected with accident liability has been found to be a peculiarity of the individual (as measured by psychophysical tests) allows us to differentiate between accident proneness and accident liability. Accident proneness is a narrower term than accident liability and means a personal idiosyncracy predisposing the individual who possesses it in a marked degree to a relatively high accident rate. Accident liability includes all the factors determining accident rate, accident proneness refers only to those that are personal."

In a series of papers, Farmer and Chambers' enthusiasm for the concept expanded until in 1939 we read: "accident proneness is no longer a theory but an established fact and must be regarded as an important element"; and in 1941 (Chambers and Yule): "it is relatively stable, in the sense that persons with a larger number of accidents than their fellows in one observational period tend to have more accidents in subsequent periods... also that persons sustaining a number of one kind of accident tend also to sustain a number of other types."

Meanwhile an important but neglected contribution to accident theory had been made by Lüders [28]. Lüders clearly understood the relationship between the two viewpoints concerning compound Poisson distributions, and he used the Poisson singlet, Poisson doublet, etc., approach to analyze data on road and rail deaths. He derived the negative binomial as a Poisson $\vee$ logarithmic distribution and considered also the Hermite distribution (the p.g.f. of which is $\exp\{a_1(s-1) + a_2(s^2 - 1)\}$), the distribution with p.g.f. equal to $\exp\{a_1(s-1) + a_2(s^2 - 1) + a_3(s^3 - 1)\}$ (which the present author has tentatively called the tri-Poisson distribution, Kemp [138]), and a convolution of the Poisson and negative binomial distributions. These were all derived as generalized Poissons; in particular, this is the first instance in the literature of the negative binomial derived in this manner, and this model did not reappear in the literature until Quenouille [51] developed it in a biometrical context.

During the same period actuarial risk theory continued to be developed in Scandinavia. Lundberg [23] produced a detailed account of his studies in German, see also Lundberg [26, 29] in English; also at Stockholm, Cramer and his associates were working in this field (see, e.g., Cramer [21], Esscher [25]. Esscher not only allowed the amount of a claim to vary, obtaining the resultant generalized Poisson distribution in a more easily recognizable form than Lundberg, but also the possibility of occurrence of a claim to vary, thus initiating the study of compound Poisson processes in actuarial risk theory. Segerdahl began his study of random processes and the ruin problem in collective risk theory with his doctoral thesis, Segerdahl [36]; however, Ove Lundberg's outstandingly important doctoral thesis was not produced until the following year.

In France, time-dependent compound Poisson processes arising from accident insurance theory were under investigation by Dubourdieu [30, 31].

5. 1940 - 1949

Ove Lundberg's [38] contribution to collective risk theory has proved extremely valuable; his thesis was published immediately as a book which has recently been reissued in paperback. In particular, Lundberg developed the theory of compound Poisson processes by introducing the idea of heterogeneity of risk conditions both in space and in time. He not only obtained the negative binomial in this manner as a compound Poisson distribution (c.f. Greenwood and Yule [11] but also derived it from the Polya process postulates, i.e., as a nonstationary pure birth process with the birth rate dependent on time. He also showed that compounding a binomial with respect to a beta (Pearson Type I) distribution yields the Polya distribution. Among the few disadvantages to the reader are lack of an index and the small number of references.

Models similar to those of Lundberg were investigated by Ammeter [46, 49], who later acknowledged Lundberg's priority. There also seems to have been a considerable delay before Lundberg's work became well-known in Western Europe and America.

During the nineteen-forties several important statistical papers on compound and generalized Poisson distributions were published, e.g., Satterthwaite [41], Feller [42], Cernuschi and Castagnetto [44], and Maceda [47], with applications as various as entomological counts, litter sizes, and match scores, as well as insurance claims. These papers, together with that of Neyman [35], are convincing evidence of the fruitfulness of ideas arising in accident and insurance studies as regards other areas of statistical investigations.

Feller [42] and Maceda [47] considered the properties of both compound and generalized Poissons; Feller brought out the distinction between true and apparent contagion. Cernuschi and Castagnetto, following Lüders; convolution of Poisson singlet, doublet, etc., method, derived the Neyman Type A distribution as a Poisson $_{\bigvee}$ Poisson, whereas Neyman [35] had derived it as a Poisson $_{\bigwedge}$ Poisson distribution. It is interesting to note that, during this period, the Poisson $_{\bigwedge}$ gamma $\sim$ negative binomial result was independently derived by several workers, e.g., Fisher [40] and Ottestad [43].

Returning now to the history of "accident proneness," Chambers and Yule's [39] paper is perhaps most noteworthy for a very good account in the discussion by Irwin of models leading to the negative binomial distribution. It is interesting to note that most workers in this field took the implicit view that not only was accident proneness an established fact, but also that the gamma was the *only* possible proneness distribution. Greenwood [54], for example, held "not only that 'accident proneness' is a human characteristic, but that it is measurable...but so far no tests, or battery of tests, has been found which would exclude the highly accident-prone *without* excluding a considerable proportion of entrants who are not specially 'bad risks'," and "a negative binomial distribution of accidents, although a necessary, was not a sufficient condition for the validity of the proneness hypothesis." It is, however, only fair to point out that Greenwood was well aware that proneness might not have a gamma distribution. He did, however, consider that if heterogeneity was present, then a negative binomial would result--this seems to be based on a belief that aggregating (in the sense of adding together the corresponding frequencies) two or more nonidentical Poissons (or negative binomials) must lead to a negative binomial distribution.

An early criticism of the usefulness of accident rate as a

measure of accident proneness came from Cobb [37], who distinguished two sets of factors influencing the susceptibility to accident of a given individual--his exposure to accidents and his personal aptitude as a driver. Furthermore, he considered the role of luck to be underestimated; from the records of nearly 30,000 drivers sustaining a total of approximately 7,000 accidents, he showed that only about a fifth of the variance remained unexplained by a simple 'luck' assumption.

Smeed [52], in his detailed examination of the statistical aspects of road safety, remarked that "perhaps more scientific papers have been written on the subject of accident proneness than on any other aspect of road safety." He too recognized that "the repetition of accidents to the same individuals found in the report may not--as is claimed--have been due to the accident proneness of the individuals concerned. It may have been at least partly due to the different liabilities to accident on the routes on which they were driving, the vehicles which they drove and the variation in mileage." Moreover he emphasized the (still serious) problem of getting reliable (uncorrupt) data on road accidents--some drivers, for example, hesitate to report their 'damage only' accidents, as opposed to their 'personal injury' accidents. An interesting feature of Smeed's paper is his investigation into the distribution of accident proneness. He obtained the variances and third moments for a standardized unit mean from the factorial moments of eleven different sets of accident data and found few signs of agreement between the figures. In one case a negative variance is inconsistent with the familiar negative binomial assumption.

6. 1950 - 1959

The big step forward concerning accident statistics during this decade came from the introduction of bivariate methods. Maritz [56], Arbous and Kerrich [57], and Bates and Neyman [64] all adopted the idea of detecting accident proneness by splitting a lengthy period of observation into two nonoverlapping periods and then correlating the numbers of accidents suffered by an individual during the two periods. This idea had, in fact, already occurred to Greenwood and Woods [10]. Under their hypothesis (a) of pure chance there should be no correlation;

for hypotheses (b) and (c) correlation would occur, but for (b) removal of individuals sustaining no accidents in the first period should increase the observed correlation, whereas this would not be so for hypothesis (c). Their data favored the latter interpretation. Newbold [17] also gave estimated correlations on the assumption of (c) with constant individual liabilities; however, there was no real attempt to consider bivariate models until the fifties.

Maritz [56] adopted a correlated bivariate Poisson model. This model implies that proneness is the same for the whole population, that the distribution within each period is Poisson, and yet that an individual's performances in the two periods are correlated. A logical difficulty arises because the distribution of accidents over the sum of the two periods is not itself Poisson (in fact, it is Hermite); hence arbitrarily splitting the total time interval produces different distributions. Maritz [56] also pointed out that a bivariate uncorrelated negative binomial model could be constructed.

Arbous and Kerrich [57] produced a major paper in which they reviewed and criticized severely the whole concept of accident proneness, pointing out that no real progress had been made since Newbold [17, 18]; they examined in detail the optimistic and, in their view, misleading conclusions drawn from data and statements made during the intervening period. Their paper has a good summary of the mathematical background and includes a new stochastic treatment of Greenwood's hypothesis (b), obtaining the 'burnt fingers' distribution from a stochastic process similar to that of McKendrick [16]. Arbous and Kerrich explicitly introduce the bivariate negative binomial distribution, deriving it on both proneness and true contagion hypotheses. The proneness derivation requires the accident distributions in the two periods to be independent Poissons, with means $\lambda\delta_0$ and $\lambda\delta_1$, and λ to have a gamma distribution. The distribution of accidents over the whole period then, is, univariate negative binomial, as are the marginal distributions. They apply their theory to examples of data involving minor and major accidents separately in two successive periods, minor and major accidents in the same period, and different types of accident in the same period--all with discouraging results. They also show that, in principle, the bivariate approach should enable one to discriminate between proneness and contagion hypotheses.

The Bates and Neyman paper [64] contains much more of general statistical interest and less of interest concerning proneness *per se* than Arbous and Kerrich. They derive the multivariate negative binomial and explore its properties in considerable detail. Their examination of the two-variable case involves considering the joint distribution of two types of accident in the same or different periods, the motivation being to use 'light' (predictor) accidents in one period to forecast 'severe' accidents in the future.

Difficulties arise if one severe accident so incapacitates an individual as to affect his future exposure to accident situations; Bates and Neyman investigated this case. They also suggested that a profitable alternative approach to the whole problem was via the distribution of inter-accident times; however, this continuous distribution approach is outside the scope of this article. It is interesting to note that, when deriving the proneness model, they considered that "the adoption of the Type III law is justified both by its flexibility as an interpolation formula and by the tradition established by Greenwood, Yule and Newbold."

For the bivariate negative binomial, the regression of one variable on the other is linear. Johnson [82] showed that, for a compound Poisson model, the gamma is the *only* compounding distribution giving rise to this property.

Adelstein [63] criticized the accident proneness concept (for the same kind of reasons as given by Arbous and Kerrich), together with its treatment in certain psychology textbooks and the use made of it in a number of practical accident studies which had appeared in the nineteen forties.

Another highly critical paper was that by Fitzpatrick [84].

A new look at univariate accident distributions was taken by Irwin [73], who demonstrated that certain distributions (in particular, the 'burnt fingers') can be obtained via their transition probabilities.

During this period the relevant areas of discrete distribution theory advanced considerably. Notable features were the development of sophisticated generalized and compound Poisson models and their empirical application (e.g., Anscombe [53], Skellam [69], Beall and Rescia [70], Evans [72], McGuire *et al*. [83], and Gurland [85]). Such papers again provide instances of the useful extension into a variety of other fields of theory and methods largely originating in the accident context. The mathematical theory of generalized Poisson distributions continued to advance--Renyi and others presented a series of papers in 1950, 1951, and 1952 to the Hungarian Mathematical Academy [55, 60, 62, 68]; and later a number of papers appeared in France and Belgium (e.g., Kaufmann and Cruon [88], Girault [87], Thyrion [90]). Note that, whereas Renyi and his colleagues used the term 'composed' to denote 'generalized,' Consael [65, 66] used it to refer to compound (Poisson distributions) in his substantial theoretical contributions.

Meanwhile, considerable emphasis was placed on the ruin problem in collective risk theory (e.g., Arfwedson [75, 76 and Segerdahl [79]), but this is essentially continuous distribution theory. A balanced appraisal of the state of the theory to date was given by Cramer [77]. Philipson [78, 80] further developed the application of stochastic processes in risk theory.

7. 1960 TO THE PRESENT

It is difficult to make any overall assessment of the position in the current decade, since development is continuing rapidly; our coverage here is somewhat sketchy.

In the actuarial field the ATSIN (actuarial studies in nonlife insurance) organization, set up after the International Actuarial Congress of 1951, held regular colloquia (see, e.g., Cramer [97], Philipson [103], de Jongh [100] and ASTIN Colloquium [95]), and began publication of the ASTIN Bulletin in 1960. Philipson has produced a series of papers devoted particularly to extending the theory of compound Poisson processes. Cramer [113] has related model building to risk theory; in a wideran ing review he deals with many topics relevant to this paper. In Borch [134] and the discussion which followed, the paucity of contact between actuarial risk theory and the mainstream of probability and mathematical statistics was commented upon.

The papers by Philipson [92] and Delaporte [98] provide a direct link between the risk and proneness theories (an earlier link is to be found in the final chapter of Lundberg [38]).

Edwards and Gurland [99] introduced an extended proneness model (rediscovered by Subrahmaniam [132]): while the bivariate negative binomial considered by earlier authors arose from compounding two independent Poissons by a gamma distribution, Edwards and Gurland removed the independence requirement. This leads to a more general form of the distribution; although the marginal distributions are negative binomial, the distribution of the sum is no longer so, but regression is still linear. Applied to one type of accident in two time periods (as proposed by the authors), this seems to suffer from the same logical difficulties as the bivariate correlated Poisson model of Maritz [56]; application to two different types of accident would be more realistic, and this is the model in Subrahmaniam's derivation. An application of the original Bates-Neyman model is given in Mellinger et al. [121].

A major innovation in road accident theory was made by Cresswell and Froggatt [106]. They rejected the concept of accident proneness and proposed the following new hypothesis to replace it: each driver is liable to have 'spells' (periods of time) within which all his accidents must occur; the numbers of spells per subject and accidents per spell have Poisson distributions (there is probably an inconsistency here) with parameters λ and θ, equal and immutable for all drivers; all drivers are equally 'exposed to risk.' They called the resultant the 'Long' distribution; it is, in fact, the Neyman Type A distribution.

Recognizing that certain types of accidents could occur to a driver outside spells, Cresswell and Froggatt introduced a random component which has a Poisson distribution, parameter ϕ. This results in another generalized Poisson, the 'Short' distribution, with p.g.f.

$$G(s) = \exp \{\lambda[\exp \theta(s-1) - 1] + \phi (s-1)\}.$$

The spells hypothesis implies zero or very low correlation between successive periods--in practice correlations have tended to be decidedly low. The book also contains much carefully collected data, details of the clinical studies carried out on the drivers involved, a general review of the proneness concept, and a very good bibliography.

Irwin [114], in a detailed review of Cresswell and Froggatt, pointed out, among other things, that since the Neyman Type A is both Poisson $\wedge$ Poisson and Poisson $\vee$ Poisson, both the 'Long' and 'Short' distributions can be given a proneness interpretation; so a good fit of either does not discredit the proneness hypothesis.

Irwin's personal view is that accident proneness "does exist and plays a modest part in accident causation."

Kemp [137] remarked that, because the negative binomial is Poisson $\vee$ logarithmic, it too can be given a 'spells' interpretation (incidentally, it is interesting to reflect that Lüders [28], having derived the negative binomial as Poisson $\vee$ logarithmic, then obtained a three-parameter distribution by convoluting it with an independent Poisson). Kemp also showed that, in general, the 'Short' distribution will be virtually uninterpretable in terms of the assumed model even if it fits the data well--this is because of the extremely high correlations between the parameter estimates due to the mathematical structure of the distribution. This problem of models, which may be perfectly meaningful physically yet have very bad statistical properties, is one which so far has not received sufficient attention outside accident studies.

Irwin [141] breaks new ground with a three-parameter distribution which enables him to partition variance into separate components due to randomness, liability and proneness. Irwin does not consider problems of estimation; and, at the time of writing, no further results have been published, so one cannot assess the viability of the model in practice.

The spells and proneness hypotheses have been applied to data on industrial absenteeism by Froggatt [116] (following on the work of Arbous and Sichel [74]) and to medical consultations by Froggatt *et al*. [145]. The latter application is of interest as a situation in which 'proneness' appears to play a major part. Froggatt has also used the Hermite distribution instead of the Neyman Type A in a 'spells' context; see also Froggatt's paper in this volume.

A number of papers have appeared in which properties of particular distributions are studied and applications to proneness theory are considered (e.g., Philipson [91] and Bhattacharya [127]). Several authors have examined the effect of removing individuals sustaining a specified number of accidents from a population where proneness is assumed to vary (e.g., Haight [117], Bhattacharya and Holla [115], and Holla [129]).

Further progress has been made with multivariate models, e.g., in Patil [123]. In the univariate area, numerous papers on properties and applications of contagious distributions have appeared; no attempt has been made to deal with these, and only a very limited selection is given in the bibliography. One should note, however, the publication of three books during the present decade, Patil [124], Haight [135] and Patil and Joshi [142]. Taken together, these contain an immense amount of information about the distributions discussed in this paper. To the best of the author's knowledge, the only important book on accident proneness published during the current decade is that of Cresswell and Froggatt [106] (Ashton's book [126] contains only a short section on accidents), while regrettably no book on the related actuarial theory has been published (although Philipson [143] states that one is under preparation by Beard, Pentikainen and Pesonen).

8. CONCLUSION

We conclude with an extract from the Industrial Fatigue Research Board's preface to Newbold [17]:

"... the Board think that the limits of knowledge that can be acquired solely by statistical methods have now been reached, and feel, indeed, that one of the main uses of the investigation has been to indicate how far statistical enquiry can go, and thereby to clear the way for the application of psychological and physiological methods to the study of the individual."

It is fortunate for statistics that the implications of this comment have been so successfully ignored.

ACKNOWLEDGEMENTS

I am deeply indebted to my wife, Dr. A. W. Kemp, and to Professor P. Froggatt for many valuable discussions on these topics.

SELECTED BIBLIOGRAPHY

[1] Bortkiewicz, L. 1898. Das Gesetz der kleinen Zahlen. Leipzig, Teubner.

[2] Erlang, A. K. 1909. Probability calculus and telephone conversations. Nyt Tidsskrift for Matematik B 20:33-9.

[3] Lundberg, F. 1909. Uber die Theorie der Rückversicherung. Trans. VI Int. Congr. Actuaries 1:877-955.

[4] Bateman, H. 1910. On the probability distribution of α particles. Phil. Mag. & J. Sci. 6th Ser. 20:704-7.

[5] Yule, G. U. 1910. On the distribution of deaths with age when the causes of death act cumulatively, and similar frequency distributions. J. Roy. Stat. Soc. 73:26-38.

[6] McKendrick, A. G. 1914. Studies on the theory of continuous probabilities with special reference to its bearing on natural phenomena of a progressive nature. Proc. Lond. Math. Soc. (2) 13:401-16.

[7] Whitaker, L. 1914. On the Poisson law of small numbers. Biometrika 10:36-71.

[8] Bortkiewicz, L. 1915. Uber die Zeitfolge zufälliger Ereignisse. Bull. Inst. Int. Statist. 20:30-111.

[9] *Timerding, H. E. 1915. Analyse des Zufalls. Braunschweig, Friedr. Vieweg und Sohn.

[10] Greenwood, M., and Woods, H. M. 1919. A report on the incidence of industrial accidents upon individuals with special reference to multiple accidents. Rep. Industr. Fatigue Res. Bd., Lond. No. 4.

[11] ————, and Yule, G. U. 1920. An inquiry into the nature of frequency distributions representative of multiple happenings with particular reference to the occurrence of multiple attacks of disease or of repeated accidents. J. Roy. Statist. Soc. 83:255-79.

[12] Eggenburger, F., and Pólya, G. 1923. Uber die Statistik verkelterer Vorgänge. Z. Angew. Math. Mech. 3:279-89.

[13] Yule, G. U. 1925. A mathematical theory of evolution based on the conclusions of Dr. J. C. Willis, F. R. S. Phil. Trans. Roy. Soc. B 213:21-87.

[14] *Cramér, H. 1926. Review of F. Lundberg. Skand. Aktuarietidskr. 7:223-

[15] Farmer, E., and Chambers, E. G. 1926. A physchological study of individual differences in accident rates. Rep. Industr. Fatigue Res. Bd., London No. 38.

[16] McKendrick, A. G. 1926. Applications of mathematics to medical problems. Proc. Edin. Math. Soc. 44:98-130.

[17] Newbold, E. M. 1926. A contribution to the study of the human factor in the causation of accidents. Rep. Industr. Fatigue Res. Bd., Lond. No. 34.

[18] ————. 1927. Practical applications of the statistics of repeated events, particularly to industrial accidents. J. Roy. Statist. Soc. 90:487-547.

[19] Eggenburger, F., and Pólya, G. 1928. Sur l'interprétation de certaines courbes de frequence. C. R. Acad. Sci. 187:870-2.

[20] Farmer, E., and Chambers, E. G. 1929. A study of personal qualities in accident proneness and proficiency. Rep. Industr. Health Res. Bd., Lond. No. 55.

[21] *Cramér, H. 1930. On the mathematical theory of risk. Stockholm, Jubilee Volume of the Försäkringsaktiebolaget Skandia.

[22] Laurin, I. 1930. An introduction into Lundberg's theory of risk. Skand. Aktuarietidskr. 13:84-111.

[23] Lundberg, F. 1930. Uber die Wahrscheinlichkeitsfunktion einer Risikenmasse. Skand. Aktuarietidskr. 13:1-83.

[24] Pólya, G. 1930. Sur quelques points de la théorie des probabilités. Ann. Inst. Henri Poincaré 1:117-61.

[25] Esscher, F. 1932. On the probability function in the collective theory of risk. Skand. Aktuarietidskr. 15:175-95.

[26] Lundberg, F. 1932. Some supplementary researches on the collective risk theory. Skand. Aktuarietidskr. 15:137-58.

[27] Farmer, E.; Chambers, E. G.; and Kirk, F. J. 1933. Tests for accident proneness. Rep. Industr. Hlth. Res. Bd., Lond., No. 68.

[28] Lüders, R. 1934. Die Statistik der seltenen Ereignisse. Biometrika 26:108-28.

[29] *Lundberg, F. 1934. On the numerical application of the collective risk theory. Stockholm, De Forenade Jubilee Volume.

[30] Dubourdieu, J. 1938a. Remarques relatives à la théorie de l'assurance-accidents. C. R. Acad. Sci. 206:303-5.

[31] ————. 1938b. Les fonctions absolument monotones et la théorie mathématique de l'assurance-accidents. C. R. Acad. Sci. 206:556-7.

[32] Muench, H. 1938. Discrete frequency distributions arising from mixtures of several single probability values. J. Amer. Statist. Assn. 33:390-8.

[33] Farmer, E., and Chambers, E. G. 1939. A study of accident proneness amongst motor drivers. Rep. Industr. Hlth. Res. Bd., Lond. No. 84.

[34] Furry, W. H. 1939. On fluctuation phenomena in the passage of high energy electrons through lead. Phys. Rev. 52:569-81.

[35] Neyman, J. 1939. On a new class of 'contagious' distributions, applicable in entomology and bacteriology. Ann. Math. Statist. 10:35-57.

[36] *Segerdahl, C. O. 1939. On homogeneous random processes and collective risk theory. Uppsala, Almqvist and Wiksells.

[37] Cobb, P. W. 1940. The limit of usefulness of accident rate as a measure of accident proneness. J. Appl. Psychol. 24:154-9.

[38] Lundberg, O. 1940. On random processes and their application to sickness and accident statistics. Uppsala, Almqvist and Wiksells.

[39] Chambers, E. G., and Yule, G. U. 1941. Theory and observation in the investigation of accident causation. J. Roy. Statist. Soc. Suppl. 7:89-101.

[40] Fisher, R. A. 1941. The negative binomial distribution. Ann. Eugen. Lond. 11:182-7.

[41] Satterthwaite, F. E. 1942. Generalized Poisson distribution. Ann. Math. Statist. 13:410-7.

[42] Feller, W. 1943. On a general class of 'contagious' distributions. Ann. Math. Statist. 14:389-400.

[43] Ottestad, P. 1944. On certain compound frequency distributions. Skand. Aktuarietidskr. 27:32-42.

[44] Cernuschi, F., and Castagnetto, L. 1946. Chains of rare events. Ann. Math. Statist. 17:53-61.

[45] Johnson, H. M. 1946. The detection and treatment of accident prone drivers. Psych. Bull. 43:489-532.

[46] Ammeter, H. 1948. A generalization of the collective theory of risk in regard to fluctuating basic probabilities. Skand. Aktuarietidskr. 31:171-98.

[47] Maceda, E. C. 1948. On the compound and generalized Poisson distributions. Ann. Math. Statist. 19:414-16.

[48] Ottestad, P. 1948. On some discrete frequency functions. Skand. Aktuarietidskr. 31:1-13.

[49] *Ammeter, H. 1949. Die Elemente der kollektiven Risikotheorie von festen und zufallsartig schwankenden Grundwahrscheinlichkeiten. Mitt. Verein. Schweiz. Versich-Math. 49:35-95.

[50] Mintz, A., and Blum, M. L. 1949. A re-examination of the accident proneness concept. J. Appl. Psych. 33:195-211.

[51] Quenouille, M. H. 1949. A relation between the logarithmic, Poisson, and negative binomial series. Biometrics 5:162-4.

[52] Smeed, R. J. 1949. Some statistical aspects of road safety research. J. Roy. Statist. Soc. A. 112:1-34.

[53] Anscombe, F. J. 1950. Sampling theory of the negative binomial and logarithmic series distributions. Biometrika 37:358-82.

[54] Greenwood, M. 1950. Accident proneness. Biometrika 37:24-9.

[55] Jánossy, L.; Rényi, A.; and Aczél, J. 1950. On composed Poisson distributions, I. Acta. Math. Acad. Sci. Hung. 1:209-24.

[56] Maritz, J. S. 1950. On the validity of inferences drawn from the fitting of Poisson and negative binomial distributions to observed accident data. Psych. Bull. 47:434-43.

[57] Arbous, A. G., and Kerrich, J. E. 1951. Accident statistics and the concept of accident proneness. Biometrics 7:340-432.

[58] Blum, M. L., and Mintz, A. 1951. Correlation versus curve fitting in research on accident proneness: reply to Maritz. Psych. Bull. 48:413-18.

[59] Burke, C. J. 1951. A chi-square test for 'proneness' in accident data. Psych. Bull. 48:496-504.

[60] Rényi, A. 1951. On composed Poisson distributions II. Acta. Math. Acad. Sci. Hung. 2:83-96.

[61] Sichel, H. S. 1951. The estimation of the parameters of a negative binomial distribution with special reference to psychological data. Psychometrika 16:107-27.

[62] Aczél, J. 1952. On composed Poisson distributions, III. Acta. Math. Acad. Sci. Hung. 3:219-23.

[63] Adelstein, A. M. 1952. Accident proneness: a criticism of the concept based upon an analysis of shunters' accidents. J. Roy. Statist. Soc. A 115:354-410.

[64] Bates, G. E., and Neyman, J. 1952. Contributions to the theory of accident proneness: I. An optimistic model of the correlation between light and severe accidents; II. True or false contagion. Univ. Calif. Publ. Statist. I 215-75.

[65] Consael, R. 1952a. Sur les processus de Poisson du type composé. Acad. Roy. de Belg. Bull. Classe des Sci. 5e Serie 38, 442-61.

[66] ————. 1952b. Sur les processus composés de Poisson à deux variables aleatoires. Acad. Roy. Belg. Classe des Sci. Memoires 27(6):4-43.

[67] Maritz, J. S. 1952. Note on a certain family of discrete distributions. Biometrika 39:196-8.

[68] Prékopa, A. 1952. On composed Poisson distributions, IV (remarks on the theory of differential processes). Acta. Math. Acad. Sci. Hung. 3:317-25.

[69] Skellam, J. G. 1952. Studies in statistical ecology. I Spatial pattern. Biometrika 39:346-62.

[70] Beall, G., and Rescia, R. R. 1953. A generalization of Neyman's contagious distributions. Biometrics 9:354-86.

[71] Bliss, C. I., and Fisher, R. A. 1953. Fitting the negative binomial distribution to biological data and note on the efficient fitting of the negative binomial. Biometrics 9:176-200.

[72] Evans, D. A. 1953. Experimental evidence concerning contagious distributions in ecology. Biometrika 40:186-211.

[73] Irwin, J. O. 1953. On the 'transition probabilities' corresponding to any accident distribution. J. Roy. Statist. Soc. B. 15:87-9.

[74] Arbous, A. G. and Sichel, H. S. 1954. New techniques for the analysis of absenteeism data. Biometrika 41:77-90.

[75] Arfwedson, G. 1954. Research in collective risk theory I. Skand. Aktuarietidskr. 37:191-223.

[76] ————. 1955. Research in collective risk theory II. Skand. Aktuarietidskr. 38:53-100.

[77] Cramer, H. 1955. Collective risk theory. Jubilee Vol. of Försäkringsaktiebolaget Skandia, Esselte, Stockholm.

[78] Philipson, C. 1955. A tentative application of the collective risk theory to crop insurance. Skand. Aktuarietidskr. 38:201-53.

[79] Segerdahl, C. O. 1955. When does ruin occur in the collective theory of risk? Skand. Aktuarietidskr. 38:22-36.

[80] Philipson, C. 1956. A note on different models of stochastic processes dealt with in the collective theory of risk. Skand. Aktuarietidskr. 39:26-37.

[81] Gurland, J. 1957. Some interrelations among compound and generalized distributions. Biometrika 44:265-8.

[82] Johnson, N. L. 1957. Uniqueness of a result in the theory of accident proneness. Biometrika 44:530-1.

[83] McGuire, J. U.; Brindley, T. A.; and Bancroft, T. A. 1957. The distribution of European corn borer larvae Pyrausta Nubilalis (Hbn.) in field corn. Biometrics 13:65-78.

[84] Fitzpatrick, R. 1958. The detection of individual differences in accident susceptibility. Biometrics 14:50-68.

[85] Gurland, J. 1958. A generalized class of contagious distributions. Biometrics 14:229-49.

[86] *Häkkinen, S. 1958. Traffic accidents and driver characteristics. Helsinki, Finland's Institute of Technology.

[87] Girault, M. 1959. Remarques sur le processus de Poisson et sur quelques généralisations. Rev. Franc. Rech. Opérationnelle 3:145-52.

[88] Kaufmann, A., and Cruon, R. 1959. Le processus de Galliher. Rev. Franc. de Rech. Opérationnelle 3:137-44.

[89] Nass, C. A. G. 1959. The χ^2 test for small expectations in contingency tables, with special reference to accidents and absenteeism. Biometrika 46:365-85.

[90] Thyrion, P. 1959. Sur une propriété des processus de Poisson généralisés. Bull. Assoc. Roy. des Act. Belges 59:35-46.

[91] Philipson, C. 1960a. The theory of confluent hypergeometric functions and its application to compound Poisson processes. Skand. Aktuarietidskr. 43:136-62.

[92] ———. 1960b. Note on the application of compound Poisson processes to sickness and accident statistics. ASTIN Bull. 1, 224-237

[93] Smeed, R. J. 1960. Proneness of drivers to road accidents. Nature Lond. 186:273-5.

[94] Thyrion, P. 1960. Note sur les distributions 'par grappes'. Bull. Assoc. Roy. Act. Belges 60:49-66.

[95] ASTIN Colloquium 1961. Abstract of discussions. Skand. Aktuarietidskr. 44:251-62.

[96] Bartko, J. J. 1961. The negative binomial distribution: a review of properties and applications. Virginia J. Sci. 12:18-37.

[97] Cramér, H. 1961. Introductory lecture: ASTIN Colloquium 1961. Skand. Aktuarietidskr. 44:137-44.

[98] Delaporte, P. J. 1961. Analyse statistique de la survenance des accidents d'automobiles et formation d'un tarif d'assurance. Bull. Inst. Int. Statist. 39(4):239-48.

[99] Edwards, C. B., and Gurland, J. 1961. A class of distributions applicable to accidents. J. Amer. Statist. Assoc. 56:503-17.

[100] Jongh, B. H. de 1961. A short note on extended Poisson processes. Skand. Aktuarietidskr. 44:210-11.

[101] Khatri, C. G., and Patel, I. R. 1961. Three classes of univariate discrete distributions. Biometrics 17:567-75.

[102] Philipson, C. 1961a. On a class of distribution functions as applied to different stochastic processes. Skand. Aktuarietidskr. 44:20-54.

[103] ———. 1961b. An extension of the models usually applied to the theory of risk. Skand. Aktuarietidskr. 44:223-39.

[104] ————. 1961c. A generalization of the stochastic processes commonly applied to the theory of casualty risk. Probability - and intensity - functions of some compound Poisson processes. Bull. Inst. Int. Stat. 39:249-60.

[105] Bohman, H., and Esscher, F. 1963. Studies in risk theory with numerical illustrations concerning distributions and stop loss premiums I. Skand. Aktuarietidskr. 34:173-225.

[106] Cresswell, W. L., and Froggatt, P. 1963. The causation of bus driver accidents. London, Oxford University Press.

[107] Philipson, C. 1963a. On Esscher transforms of distribution functions defining a compound Poisson process for large values of the parameter. Skand. Aktuarietidskr. 34:226-36.

[108] Philipson, C. 1963b. On the difference between the concepts 'compound' and 'composed' Poisson processes. ASTIN Bull. 2:445-51.

[109] Philipson, C. 1963c. A note on s. or n.s. cPp. Skand. Aktuarietidskr. 34:239-42.

[110] ————. 1963d. Quelques processus applicables dans l'assurance et dans la biologie. Bull. Assoc. Roy. Act. Belg. 61:51-65.

[111] Thyrion, P. 1963. Note sur une transformation des distributions generalisées. Bull. Assoc. Roy. Act. Belg. 61:67-79.

[112] Tucker, H. G. 1963. An estimate of the compounding distribution of a compound Poisson distribution. Theory of Prob. & its Appl. 8:195-200.

[113] Cramér, H. 1964. Model building with the aid of stochastic processes. Technometrics 6:133-59.

[114] Irwin, J. O. 1964. The personal factor in accidents. J. Roy. Statist. Soc. A 127:438-51.

[115] Bhattacharya, S. K., and Holla, M. S. 1965. On a discrete distribution with special reference to the theory of accident proneness. J. Amer. Statist. Assoc. 60:1060-66.

[116] Froggatt, P. 1965. One-day absence in industry. J. Stat. Soc. Inq. Soc. Ireland 21(3):166-78.

[117] Haight, F. A. 1965. On the effect of removing persons with N or more accidents from an accident prone population. Biometrika 52:298-300.

[118] Holla, M. S., and Bhattacharya, S. K. 1965. On a discrete compound distribution. Ann. Inst. Statist. Math. Tokyo 15:377-84.

[119] Kemp, C. D., and Kemp, A. W. 1965. Some properties of the 'Hermite' distribution. Biometrika 52:381-94.

[120] Martin, D. C., and Katti, S. K. 1965. Fitting of some contagious distributions to some available data by the maximum likelihood method. Biometrics 21:34-48.

[121] Mellinger, G. D., et al. 1965. A mathematical model with applications to a study of accident repeatedness among children. J. Amer. Statist. Assoc. 60:1046-59.

[122] Neyman, J. 1965. Certain chance mechanisms involving discrete distributions. Sankhyā A 27:249-58.

[123] Patil, G. P. 1965a. Certain characteristic properties of multivariate discrete probability distributions akin to the Bates-Neyman model in the theory of accident proneness. Sankhya A 27:259-70.

[124] Patil, G. P. (Ed.) 1965b). Proc. Int. Symp. Classical and Contagious Discrete Distributions, Montreal. Calcutta, Statistical Publishing Society.

[125] Philipson, C. 1965. A generalized model for the risk process and its application to a tentative evaluation of outstanding liabilities. ASTIN Bull. 3:215-38.

[126] Ashton, W. D. 1966. The theory of road traffic flow. London: Methuen.

[127] Bhattacharya, S. K. 1966. Confluent hypergeometric distributions of discrete and continuous type with applications to accident proneness. Calcutta Statist. Assoc. Bull. 15:20-31.

[128] Holgate, P. 1966. Bivariate generalizations of Neyman's type A distribution. Biometrika 53:241-5.

[129] Holla, M. S. 1966. On a Poisson-inverse gaussian distribution. Metrika 11:115-21.

[130] Katti, S. K. 1966. Interrelations among generalized distributions and their components. Biometrics 22:44-52.

[131] Kemp, A. W., and Kemp, C. D. 1966. An alternative derivation of the Hermite distribution. Biometrika 53:627-8.

[132] Subrahmaniam, K. 1966. A test for 'intrinsic correlation' in the theory of accident proneness. J. Roy. Statist. Soc. B 28:180-89.

[133] Bhattacharya, S. K. 1967. A result on accident proneness. Biometrika 54:324-5.

[134] Borch, K. 1967. The theory of risk. J. Roy. Statist. Soc. B 29:432-52.

[135] Haight, F. A. 1967. Handbook of the Poisson distribution. New York: Wiley.

[136] Johnson, N. L. 1967. Note on a uniqueness relation in certain accident proneness models. J. Amer. Statist. Assoc. 62:288-9.

[137] Kemp, C. D. 1967a. On a contagious distribution suggested for accident data. Biometrics 23:241-55.

[138] Kemp, C. D. 1967b. 'Stuttering-Poisson' distributions. J. Stat. Soc. Inq. Soc. Ireland 21(5):151-157.

[139] Kemp, C. D., and Kemp, A. W. 1967. A special case of Fisher's 'modified Poisson series'. Sankhyā A 29:103-4.

[140] Bhattacharya, S. K. 1968. Bayes approach to compound distributions arising from truncated mixing densities. Ann. Inst. Statist. Math. Tokyo 20:375-81.

[141] Irwin, J. O. 1968. The generalized Waring distribution applied to accident theory. J. Roy. Stat. Soc. A 131:205-25.

[142] Patil, G. P., and Joshi, S. W. 1968. A dictionary and bibliography of discrete distributions. Edinburgh, Oliver and Boyd.

[143] Philipson, C. 1968a. A review of the collective theory of risk. I Comments on the development of the theory. Skand. Aktuarietidskr. 51:45-68.

[144] Philipson, C. 1968b. A review of the collective theory of risk. II. Skand. Aktuarietidskr. 51:117-33.

[145] Froggatt, P.; and Dudgeon, M. Y.; and Merrett, J. D. 1969. Consultations in general practice. Brit. J. Prev. Soc. Med. 23:1-11.

ON THE MEASUREMENT OF TEMPERO-SPATIAL CLUSTERING OF DISEASE: MODIFICATIONS OF EDERER-MYERS-MANTEL MODEL

K. SUBRAHMANIAM
PAUL T. MA
University of Manitoba

SUMMARY

In this paper we consider some modifications of the methods of measurement of clustering suggested by Ederer, et al. [4]. A new measure of clustering is proposed which takes into consideration the occupancy of neighboring cells as well as the multiple occupancies of a high cell. Tables of the probability distributions of the new statistics, along with the conditional expected values and conditional variances, are presented.

1. INTRODUCTION AND REVIEW

It is of basic interest to an epidemiologist to determine the etiology of a disease before any remedial measures can be suggested. For diseases such as cancer, for which the causative factor is still a matter for conjecture, the mode of transport of the disease should be studied before any ideas regarding the etiology can be formulated. A number of diseases of equally rare occurrence are also under investigation in this respect.

One important factor in the epidemicity of a disease is the so-called "clustering" of its incidence over space and over time. In the case of a disease which is not widespread, it is more logical to consider a possible interaction between the incidence over time and space. This idea of an interaction (or the tempero-spatial clustering of disease) in leukemia was first considered by Kellet [6]. He was led to this possibility by his analysis of 63 leukemia cases in the Royal Victoria Infirmary over the period 1932-36. Subsequently, a number of empirical studies have been undertaken to confirm his ideas.

Knox [7, 8] has suggested a method in which he examines the correlation between short geographic distances and short time intervals. This is done by looking at all the $\binom{n}{2}$ "pairs" that can be formed from the "n" incident cases. A table is constructed as follows:

		SPACE	
		Adjacent	Not Adjacent
TIME	Adjacent	X	
	Not Adjacent		
			$\binom{n}{2}$

The entry "X" in the top left-hand cell is assumed to be a Poisson-distributed random variable. A graph theoretic verification of this assumption on the distribution of X has been made by Barton and David [2]. Around the same time, Pike [16] verified empirically, for cases studied by Knox, that X does have a Poisson distribution. More recently Pike and Smith [17] have considered a generalization of Knox's approach to time-space clustering.

An alternative method involving the F-test has been suggested by David and Barton [3]. The procedure is based on the analysis of the total distances between the $\binom{n}{2}$ pairs of points in the plane subdivided into the distances between subgroups and within subgroups of points. An analysis-of-variance test with the usual normal approximations is recommended. As applied to the test of epidemicity, one has to define clusters in time and space before using this procedure.

A method due to Lundin _et al_. [9] pairs observed sets of cases into geographically proximate and temporally related subunits by means of a complicated system of criteria. These criteria use an index case as the basic unit, and subsequent cases are related to this index case. A comparison is made of the observed configuration with that expected with a random allocation by using a χ^2 test of significance.

One feature that is common to all the methods discussed is the analysis of the discrepancy between observed and expected frequencies of "paired cases" in a predefined unit of space and time. Fortunately, it has been demonstrated by Barton et al. [1] that the choice of units does not affect the results to any great extent, provided a choice is made before the data themselves are looked at.

Ederer _et al_. [4] have preferred another approach which does not use the paired-cases technique. It is based on the analysis of the observed defined clusters and the expected ones on the assumption of no clustering. We will discuss this method at a greater length in the next section.

Mustacchi _et al_. [14] have made a comparison of various methods, applying them to the distribution of fatal poliomyelitis in San Francisco over the ten year period 1946-1955, and found them to be consistent.

A discussion of these and other methods of studying clustering of disease is given by Mantel [11]. In a recent thesis by the co-author, Ma [10], some applications have been suggested for these methods.

2. EDERER-MYERS-MANTEL MODEL

Consider the problem of placing r balls in n cells, all the balls and the cells being distinguishable. If we have the configuration in which the occupancy numbers are $\nu_1, \nu_2, \ldots, \nu_n$, _not necessarily in that order_, with the frequencies $j_0, j_1, \ldots, j_r$

for the occupancy numbers $0,1,2,\ldots,r$, then

$$\sum_{j=1}^{n} \nu_j = r \qquad \text{and} \qquad \sum_{k=0}^{r} j_k = n.$$

The probability of such an arrangement is

$$P[\nu_1,\ldots,\nu_n;\ j_0,\ldots,j_r] = \frac{n!}{j_0!\ j_1!\ \ldots\ j_r!}\ \frac{r!}{\nu_1!\ \nu_2!\ \ldots\ \nu_n!}\ \frac{1}{n^r}. \quad (1)$$

A situation of this type has been very well exemplified by Feller ([5], pp. 37-8). It should be noted that the statement by Ederer *et al.* to the effect that the probability in (1) is that of an ordered set is not quite true.

As pointed out by Ederer *et al.*, there is an analogy between the occupancy model and their method for the analysis of disease clustering. In their development the authors divided a given geographic area (state of Connecticut) into k subunits (towns) $c_1,c_2,\ldots,c_k$. The total period of time, 1945-59, is divided into five-year periods so that the trend is minimized, and thus a five-year-town unit is obtained.

In their case k = 169, and hence they have 507-town-x-five-year units. A classification is made of these 507 units in accordance with the total number of "r" cases observed within such a unit. Their study is confined to the units in which there were at least two cases. The time-space unit is obtained by considering the five one-year periods within the town-x-five-year unit. An analysis is made of the "occurrence" and the "arrangement" of the r cases over these 5 time-space units. The main criteria of clustering suggested by Ederer *et al.* are

(a) m_1 = maximum number of cases occurring in a time interval (time-space unit); and

(b) m_2 = maximum number of cases occurring in two successive units.

The quantities m_1 and m_2 are referred to, respectively, as the "empiric cluster" and "two subinterval empiric cluster." Extensions to m_2 over more subintervals, as well as the possible study of vacuities (when r is large), are mentioned. Probability distributions of m_1 and m_2 are presented for selected values of r and N, the time period (i.e., the number of time-space units).

The relation between the occupancy problem and the problem of the occurrence and arrangement of the r incidences over the N subunits of time in a geographic subunit is quite obvious.

The procedure adopted for testing if the observed distribution represents a clustering of disease in space and time is to look at the distribution of the quantity

$$\bar{m}_1 = \sum_r \sum_x x\, \hat{f}(x|r), \tag{2}$$

where x is the value of m_1 and $\hat{f}(x|r)$ is the conditional frequency of x observed for a given r. If we represent by

$$p(x|r) = \text{Prob [maximum occupancy number in any \underline{one} cell over the N cells is x given that the total over the N cells is r]}$$

and we assume that f(r) the marginal frequency (observed) or r is fixed, then

$$E[\bar{m}_1] = \sum_r \sum_x x\, f(r)\, p(x|r),$$

since

$$E[\hat{f}(x|r] = f(r)\, p(x|r).$$

Hence, if we define the mean and the variance of the conditional distribution of $\bar{m}_1$ for given r (in the population) as

$$\mu(r) = \sum_x x\, p(x|r)$$

and

$$\sigma^2(r) = \sum_x [x-\mu(r)]^2\, p(x|r),$$

then

$$E[\bar{m}_1] = \sum_r \mu(r)\, f(r) \tag{3}$$

and

$$\text{Var}[\bar{m}_1] = \sum_r \sum_x x^2 f(r)\, p(x|r)\, [1 - p(x|r)],$$

which can be written as

$$\text{Var}[\bar{m}_1] = \sum_r f(r)\, \sigma^2(r). \tag{4}$$

The asymptotic distribution of $\bar{m}_1$, therefore, can be taken to be

$$\bar{m}_1 \approx N\, [\sum_r \mu(r)\, f(r),\ \sum_r f(r)\, \sigma^2(r)]. \tag{5}$$

Thus the test of significance suggested is to compare

$$\chi^2 = \frac{[\,|\bar{m}_1 - \sum_r \mu(r)\, f(r)|\ - \frac{1}{2}]^2}{\sum_r f(r)\, \sigma^2(r)}$$

with the tabulated value of chi-square with 1 degree of freedom. These developments have not been explained clearly by Ederer et al. Also their terminology in the definition of E (m_1) and Var (m_1) is not quite appropriate. They, of course, are referring to the conditional expected value, $\mu(r)$, and the conditional variance, $\sigma^2(r)$. A similar procedure can be developed for the statistic m_2.

3. MODIFICATIONS

The Ederer et al. model represents a very interesting breakaway from the standard "paired cases" techniques of Knox and the others. Its main merits lie in the simplicity of applicability in contrast to the methods involving pairing of cases. There are a number of questions that could be raised in connection with the definitions of clustering adopted by Ederer et al. In our opinion, there is a wealth of possible theoretical study to be undertaken in this regard. We shall direct our discussion to the definition of the criteria m_1 and m_2, which we consider as being open to further debate. The moot point is whether it is realistic to ignore the contiguity of occupied cells, as has been done in their definitions. The fact that contiguous cells are occupied should play a significant role in the definition of any criteria of clustering. Whereas it is reasonable to assume that the multiple occupancy of a single cell should be of primary consideration, we should not ignore contiguity of single or multiple occupancies. It is with this in mind that we suggest some modifications of m_1 and m_2. More important to our study is the definition of a new measure, which assigns different weights to the cells in deference to their occupancy numbers and takes into consideration the contiguities of these occupied cells. It appears to us that, in a disease of rare occurrence, such a consideration cannot be overstressed.

In the following we present some modifications of the Ederer et al. measures of clustering, as well as a new measure:

(i) m_1^*: In a disease of rare occurrence it is obvious that the maximal frequency occurs at $r = 2$. The definition of m_1, the maximum cell occupancy in the N cells, does not distinguish between the pattern ($r=2$, $m_1=1$), in which contiguous cells are occupied and the case in which an occupied cell is surrounded by empty cells. As has been pointed out, the latter case should be treated distinctly, and with this in mind we define:

m_1^* = maximum number of cases in a single cell, if this number is greater than or equal to 2; or
= 1 if contiguous cell occupancy numbers are 1; or
= 0 otherwise.

This definition is of significance when N is large and the disease of rare occurrence.

(ii) m^{**}: The alternative to m_1 suggested by Ederer _et al._ is to find the total occupancy number for a preassigned number t of the cells. The definition suffers from the same drawbacks as m_1, except to a larger measure. A more appropriate statistic would be to consider:

m^{**} = maximum consecutive sum of _all_ occupied cells between pairs of vacuities in the N cells, provided that this sum is greater than 1: or
= 0 otherwise.

(iii) m_w: It is quite obvious that m^{**} would tend to overestimate the effects of clustering in as much as its value does not take into account the number of cells contributing to it. This situation is corrected by attaching unequal weights to the cell contents. These weights are determined by (a) the occupancy number of the cell and (b) the condition of consistency in that the same patterns yield the same value of the statistic for different N. Thus

m_w = maximum sum of the weights in the contiguous non-zero cells with the summation over all such cells between pairs of vacuities. Individual cell weights are defined as below:

Occupancy Number	Weight per Occupant	Cell Weight
1*	0.5	0.5
2	1.0	2.0
3	1.5	4.5
4	2.0	8.0
5	2.5	12.5
⋮	⋮	⋮

*Only in the case when the cell has at lease _one_ non-zero neighbor. Otherwise, the weight is zero.

Although the value of m_w is affected by the multiplication of the cell weights by a constant, the final value of chi-square is unaffected by such an algebraic manipulation, provided that the relative weights are unchanged. It is useful to consider an example with N=3 and r=3, giving the following values of the various suggested statistics:

	m_1	m_2	m_1^*	m^{**}	m_w
3 0 0	3	3	3	3	4.5
2 1 0	2	3	3	3	2.5
2 0 1	2	2	2	2	2.0
1 1 1	1	2	1	3	1.5

It can be seen from this table that the weighting system tends to assign higher weights to multiple occupancies of cells. We feel, however, that this in no way detracts from the merits of the method. The low probabilities associated with such configurations, and also the low observed frequencies, will dampen the effects of any such overweighting. Also this scheme seems to be more appropriate than any other in the case when the incidences are spread over a number of contiguous cells, as in the last three configurations given above. These are the more relevant arrangements, especially for a disease of rare occurrence.

4. PROBABILITY DISTRIBUTION AND TESTS

Pending further investigations into more appropriate methods of testing for the significance of observed clustering, we apply the test suggested by Ederer *et al*:

$$\chi^2 = \frac{[\sum_r \sum_x x \hat{f}(x|r) - \sum_r f(r)\ E(x|r)]^2}{[\sum_r f(r)\ \mathrm{Var}\ (x|r)]},$$

with

x = statistic under consideration;
$\hat{f}(x|r)$ = observed frequency of x for a given r;
$f(r)$ = marginal frequency of r;

and $E\ (x|r)$ and $\mathrm{Var}\ (x\ r)$ as the conditional mean and conditional variance, respectively, of x given r.

Following the line of argument given in section 2, it can be readily shown that the statistic χ^2 is asymptotically distributed as a single-degree-of-freedom chi-square variable. A correction for continuity would most probably lead to more accurate results. The quantities E $(x|r)$ and Var $(x|r)$ are determined by the formulae

$$E\ (x|r) = \sum_x x\ p(x|r) \quad \text{and}$$

$$\text{Var}\ (x|r) = \sum_x [x - E(x|r)]^2\ p(x|r),$$

in which $p(x|r)$ is the conditional probability of x for given r. These formulae have been used in section 2.

Ederer et al. have tabulated the conditional probability distributions of m_1 and m_2 for selected values of r and N. Since it does not appear very fruitful to present the probability distributions extensively, in the Appendix we have tabulated the conditional probability distributions of the statistics m_1^*, m^{**} and m_w for only small values of N. We have presented the conditional moments of the statistics m_1^*, m^{**} and m_w for selected values of r and N in Table 1.

Table 1. Conditional Moments $E[x|r]$ and Var $[x|r]$ of the Statistics m_1^*, m^{**} and m_w

	E[x|r]			Var[x|r]		
N = 3	m_1^*	m^{**}	m_w	m_1^*	m^{**}	m_w
r = 2	1.1111	1.5556	1.1111	0.5433	0.6913	0.5432
3	1.8889	2.5556	2.3889	0.3210	0.2467	0.7099
4	2.3704	3.7531	3.8025	0.3072	0.3340	1.5165
5	2.7778	4.7943	5.6481	0.4938	0.3276	2.1997
N = 4						
r = 2	0.8750	1.2500	0.8750	0.6093	0.9375	0.6094
3	1.6875	2.5312	2.0156	0.3398	0.2490	0.6951
4	2.1250	3.3906	3.0781	0.3283	0.5661	1.4158
5	2.4805	4.4629	4.5263	0.3901	0.5611	2.1665
N = 5						
r = 2	0.8800	1.0400	0.7200	0.5056	0.9984	0.6016
3	1.5120	2.2800	1.7400	0.4259	0.4896	0.7464
4	1.9520	3.1936	2.6816	0.3500	0.4633	1.2058
5	2.2862	4.0304	3.7944	0.3548	0.7783	1.9925

5. DISCUSSION

It appears rather confusing to be presented with so many alternative methods of analyzing the same body of data. In a problem such as the one at hand, it is not surprising that some subjective evaluation of the various methods must be carried out and the choice of one particular procedure must be made. Although no attempt at any specific recommendation as to such a choice has been made here, we would like to point out that the method of weights is, to our minds, the best one attempted at the time of this writing. It stands to reason that any measure of clustering should make adequate correction for the occupancy of neighboring cells while attaching most significance to multiple occupancy of a single cell. Such an adjustment is provided for by the weights presented by us in section 4. Some applications of this method are being made, and we would hope to be able to better assess the relative merits of the various measures at a future date.

ACKNOWLEDGEMENTS

The authors would like to thank Dr. N. W. Choi of the University of Manitoba for bringing the problem of clustering to the attention of the second author. We owe special thanks to Dr. Piero Mustacchi for taking the trouble of supplying the raw data collected by him at the San Francisco Medical Center of the University of California. This in itself was ample encouragement. Unfortunately due to the pressures of time, we have not been able to include these data in our paper.

REFERENCES

[1] Barton, D. E., David, F. N., and Merrington, M., 1965. A criterion for testing contagion in time and space. Ann. Hum. Gen. 29:97-102.

[2] ———, ———. 1966. Random intersection of two graphs. In Research Papers in Statistics. Ed. F. N. David. 455-9. John Wiley, New York.

[3] David, F. N. and Barton, D. E. 1966. Two space-time interaction tests for epidemicity. Brit. J. Prev. and Social Medicine 20:44-8.

[4] Ederer, F.; Myers, M. H.; and Mantel, N. 1964. A statistical problem in space and time: Do leukemia cases come in clusters? Biometrics 20:626-38.

[5] Feller, W. 1957. An Introduction to Probability Theory and its Applications, Vol. I. John Wiley, New York.

[6] Kellett, C. E. 1937. Acute myeloid leukemia in one of identical twins. Arch. Dis. Child. 12:239-52.

[7] Knox, G. 1959. Secular pattern of congenital oesophageal atresia. Brit. J. Prev. and Social Medicine 13:222-6.

[8] ———. 1963. Detection of low intensity epidemicity. Brit. J. Prev. and Social Medicine 17:121-7.

[9] Lundin, F. E.; Fraumein, J. F.; Lloyd, W.; and Smith, E. M. 1966. Temporal relationships of leukemia and lymphoma deaths in neighborhoods. J. Nat. Cancer Inst. 37:123-33.

[10] Ma, Paul T. 1969. Measurement of tempero-spatial clustering of disease. Unpublished thesis presented to the Graduate Faculty of the University of Manitoba, Winnipeg, Manitoba.

[11] Mantel, N. 1967. The detection of disease clustering and a generalized regression approach. Cancer Research 22(1):209-20.

[12] ———, and Haenzel, W. 1959. Statistical aspects of the analysis of data from retrospective studies of disease. J. Nat. Cancer Inst. 22:719-48.

[13] Mustacchi, P. 1965. Some intra-city variations of leukemia incidence in San Francisco, Cancer 18:362-8.

[14] ———; David, F. N.; and Fix, E. 1965. Three tests for space-time interaction: a comparative evaluation. Proc. Fifth Berkeley Symp. on Math. Stat. and Prob., Vol. IV, 229-35.

[15] Myers, M. H.; Eisenberg, H.; and Campbell, P. G. 1965. Tempero-spatial distribution of leukemia and lymphoma in Connecticut. J. Nat. Cancer Inst. 35:625-9.

[16] Pike, M. C. 1966. Referred to as unpublished report. See [10], above.

[17] ———, and Smith, P. G. 1968. Disease clustering: a generalization of Knox's approach to the detection of space-time interactions. Biometrics 24:541-56.

[18] Pinkel, D. and Nefzger, D. 1959. Some epidemiological features of childhood leukemia. Cancer 12:351-8.

APPENDIX

In this section we tabulate the probability distributions of the statistics m_1^*, m^{**} and m_w for selected values of r and N.

Table 2. Conditional Probability Distribution of m_1^*

	m_1^*					
N = 3	0	1	2	3	4	5
r = 2	0.2222	0.4444	0.3333	-	-	-
3	-	0.2222	0.6667	0.1111	-	-
4	-	-	0.6667	0.2963	0.0370	-
5	-	-	0.3704	0.4938	0.1235	0.0123
N = 4						
r = 2	0.3750	0.3750	0.2500	-	-	-
3	-	0.3750	0.5625	0.0625	-	-
4	-	0.3750	0.7031	0.1875	0.0156	-
5	-	-	0.5859	0.3516	0.0586	0.0039
N = 5						
r = 2	0.4800	0.3200	0.2000	-	-	-
3	0.0480	0.4320	0.4800	0.0400	-	-
4	-	0.1920	0.6720	0.1280	0.0080	-
5	-	0.0384	0.6720	0.2560	0.0320	0.0016
N = 6						
r = 2	0.5556	0.2778	0.1667	-	-	-
3	0.1667	0.3889	0.4167	0.0278	-	-
4	-	0.2778	0.6250	0.0926	0.0046	-
5	-	0.0926	0.6944	0.1929	0.0193	0.0007

Table 3. Conditional Probability Distribution of m**

	m**				
N = 3	0	2	3	4	5
r = 2	0.2222	0.7778	-	-	-
3	-	0.4444	0.5556	-	-
4	-	0.0741	0.0987	0.8272	-
5	-	-	0.0823	0.0412	0.8766
N = 4					
r = 2	0.3750	0.6250	-	-	-
3	-	0.4688	0.5312	-	-
4	-	0.1641	0.2812	0.5547	-
5	-	-	0.1562	0.2246	0.6191
N = 5					
r = 2	0.4800	0.5200	-	-	-
3	0.0480	0.5760	0.3760	-	-
4	-	0.1536	0.4992	0.3472	-
5	-	0.0288	0.2880	0.3072	0.3760
N = 6					
r = 2	0.5556	0.4444	-	-	-
3	0.0556	0.6806	0.2639	-	-
4	-	0.3843	0.4012	0.2145	-
5	-	0.0926	0.3961	0.2906	0.2207

Table 4. Conditional Probability Distribution of m_w

	N = 3				N = 4				N = 5			
m_w	r=2	r=3	r=4	r=5	r=2	r=3	r=4	r=5	r=2	r=3	r=4	r=5
0.0	0.2222	-	-	-	0.3750	-	-	-	0.4800	0.0480	-	-
0.5	-	-	-	-	-	-	-	-	-	-	-	-
1.0	0.4444	-	-	-	0.3750	0.1875	-	-	0.3200	0.2880	0.0384	-
1.5	-	0.2222	-	-	-	0.1875	-	-	-	0.1440	0.0768	-
2.0	0.3333	0.2222	0.4444	-	0.2500	0.2812	0.3515	-	0.2000	0.2880	0.3072	0.0672
2.5	-	0.4444	-	-	-	0.2812	0.0938	0.1172	-	0.1920	0.2304	0.2304
3.0	-	-	0.0741	-	-	-	0.2812	-	-	-	0.1728	0.1152
3.5	-	-	-	-	-	-	-	0.2344	-	-	-	0.1536
4.0	-	-	0.1481	-	-	-	0.0703	0.0586	-	-	0.0384	0.0576
4.5	-	0.1111	0.0988	0.4527	-	0.0625	0.0938	0.2734	-	0.0400	0.0768	0.1824
5.0	-	-	0.1975	-	-	-	0.0938	0.0781	-	-	0.0512	0.0768
5.5	-	-	-	0.2469	-	-	-	0.1172	-	-	-	0.0576
6.5	-	-	-	0.1646	-	-	-	0.0585	-	-	-	0.0256
7.5	-	-	-	-	-	-	-	-	-	-	-	-
8.0	-	-	0.0370	0.0412	-	-	0.0156	0.0293	-	-	0.0080	0.0192
8.5	-	-	-	0.0823	-	-	-	0.0293	-	-	-	0.0128
12.5	-	-	-	0.0123	-	-	-	0.0039	-	-	-	0.0016

BINOMIAL ERROR MODELS IN MENTAL TESTING*

FREDERIC M. LORD
Eductional Testing Service
Princeton, New Jersey

SUMMARY

A general model for the relation between fallible measurements and "true" measurements is outlined. This model is appropriate in many scientific areas. It leads to the empirical Bayes problem of estimating the frequency distribution of true measurements for a population of measured objects. A special case important in mental test theory is considered. Practical results obtained by one estimation procedure are briefly outlined.

*This work was supported in part by contract N00014-69-C-0017 between the Office of Naval Research and Eductional Testing Service. Reproduction, translation, use and disposal in whole or in part by or for the United States Government is permitted.

Although this paper is primarily concerned with the use of binomial error models in mental testing, similar models play an important role in many scientific areas. For this reason, I will start discussing a more general model, later specializing this to the particular problems of mental test theory.

Suppose that we have many randomly chosen objects and that we make just one measurement on each object. Note that this situation is different from those considered in the classical theory of errors, which deals with repeated measurements on each object. Observed measurements are here denoted by X; the frequency distribution of X in the population of objects is written $\phi(x)$. Each observed measurement contains an error of measurement, denoted by E. This automatically implies that the difference X - E represents a true measurement

$$\xi = X - E \ . \tag{1}$$

The frequency distribution of true measurements in the population of objects is $g(\xi)$. We will assume that the errors of measurement are unbiased, so that

$$\varepsilon(E|\xi) = 0 \tag{2}$$

for every ξ. It follows that

$$\varepsilon(X|\xi) = \xi \ . \tag{3}$$

Denote by $h(x|\xi)$ the conditional distribution of x when ξ is fixed. Without further assumptions, the bivariate distribution of x and ξ is given by

$$\Psi(x,\xi) = h(x|\xi)\ g(\xi) \ ; \tag{4}$$

also, if the integral sign is properly interpreted, the marginal distribution of observed measurements is given by

$$\phi(x) = \int_{-\infty}^{\infty} h(x|\xi)\ g(\xi)\ d\xi \ . \tag{5}$$

The last equation provides the basic mathematical model with which we are concerned here. For the model to be of use, it is necessary that we know the mathematical form of h. Because of the simple relation between X, E, and ξ, $h(x|\xi)$ is the same except for a translation of origin as the conditional distribution of E for fixed ξ. In many cases we may be willing to assume that we know the form of this conditional distribution. If not, we can try various assumed forms, in each case checking the predictions made from the model (5) against actual data, until eventually we find some assumed h that, in most cases, provides correct predictions of important characteristics of real data.

If h and ϕ are known, we can use equation (5) to try to determine $g(\xi)$, the distribution of true measurements in the population. In this case, (5) is a Fredholm integral equation of the first kind. If a solution exists, methods are known for finding $g(\xi)$ satisfying the equation. Such solutions may or may not be unique. The problem of finding such solutions and determining whether or not they are unique is what I here call the mathematical problem. This type of problem is fairly well solved. In particular, there are some very elegant results by the British astronomer Eddington, and others, for the case where h is a normal distribution with mean ξ and a known constant variance (Eddington [4], Kurth [9], Pollard [16], and Trumpler and Weaver [26], sections 1.43 and 1.51). In this case, ϕ is the convolution of h and g.

Important problems arise because in practical work we never know $\phi(x)$; all we can do is observe the frequency distribution f(x) for a sample of objects drawn from $\phi(x)$. The problem of inferring $g(\xi)$ when only f(x), not $\phi(x)$, is available, I call the statistical inference problem. It is an empirical Bayes problem. Robbins [19], Wolfowitz [30], Maritz [13], Deely and Kruse [3], and others have worked on such problems.

The really serious difficulties encountered in work on the statistical inference problem arise from the fact that this problem belongs to the class of ill-posed problems, which means that small changes in f(x) may correspond to indefinitely large changes in $g(\xi)$. Recently, many good methods have been developed for solving ill-posed problems (see [1], [2], [5]-[8], [12], [14], [15], [17], [18], [20], [21]-[25], and [27]-[29]); unfortunately, these are concerned with the mathematical problem, not with the statistical inference problem. It seems likely that good methods for solving the statistical inference problem will be devised along similar lines. The method described in the present paper is developed somewhat differently.

Let me now specialize this general model to a situation of basic interest in mental test theory. Here, the "objects to be measured" are the examinees, the people to whom the test is administered. The observed measurement X is taken as the number of questions answered correctly; thus X is an integer 0,1,...,n, where n is the number of questions in the test.

The idea that the observed score X contains an error of measurement may not seem obvious to everyone who has taken a test, but it is basic to most work in mental test theory. The reason is that we usually can build many "interchangeable" forms of any particular test, each of which seems as suitable and satisfactory for our purposes as the next; and we know that examinees are likely to obtain substantially different scores on these apparently interchangeable test forms. How is the psychologist to deal in theory or in practice with such a variety of scores on the same test for a single examinee? He feels that, for most purposes, the differences between these scores are of little interest and that the thing of interest is something common to all of them. Thus it is helpful--indeed almost essential--to think that each examinee has a "true" standing on the trait measured and that any particular observed score differs from this true value by an error of measurement.

It is convenient here to define

$$\zeta = \xi/n \ , \tag{6}$$

which will also be called a true score, without creating serious confusion. Note that

$$\varepsilon \frac{X}{n} = \zeta \ ; \tag{7}$$

thus, ζ may be thought of as representing the proportion of items that the examinee could answer correctly out of a whole population of test items. Note that ζ is a continuous variable with $0 \leq \zeta \leq 1$.

In this connection, as a first approximation it is reasonable to think that the conditional distribution of observed score for fixed true score is the binomial distribution

$$h(x|\zeta) = \binom{n}{x}\zeta^x(1 - \zeta)^{n-x} \ . \tag{8}$$

Actually, a more reasonable assumption is that $h(x|\zeta)$ is a compound binomial distribution, expressible as a weighted sum of simple binomials. Although this greatly complicates any practical work, it does not affect the theory in any significant way; so the discussion here can be restricted to the simple binomial without great loss.

The model now becomes

$$\phi(x) = \int_0^1 h(x|\zeta)\ g(\zeta)\ d\zeta \ . \tag{9}$$

If we have tested a large group of examinees, we can obtain f(x) and thus approximate $\phi(x)$. The basic problem is to estimate $g(\zeta)$ from f(x).

If $g(\zeta)$ can be estimated, by $\hat{g}(\zeta)$, say, many results of practical value follow. First of all, since $h(x|\zeta)$ is assumed known, we can construct from $\hat{g}(\zeta)$ an estimate of the bivariate distribution of true score and observed score, using equation (4). If desired one can use this bivariate distribution to estimate the true score of any randomly chosen examinee from his observed score, using the regression of ζ on x, or by other procedures. Presumably, this bivariate distribution of ζ and X contains all the information that could be asked for about the relation between observed measurements and true measurements. It is the obtaining and the utilization of this information that is just what mental test theory is all about.

Although the problem of solving (9) to determine $g(\zeta)$ from $\phi(x)$ is here called the "mathematical problem," the nature of the problem should be especially clear to statisticians, as the following treatment will show. When $h(x|\zeta)$ is a binomial or compound binomial, the first n moments of $g(\zeta)$ are determined by the first n moments of $\phi(x)$. Any $g(\zeta)$ having the appropriate moments up through order n is a solution to equation (9). Thus, in the binomial case, the true-score distribution cannot be uniquely determined from the observed-score distribution, even in the entire population.

In spite of this, the problem of estimating $g(\zeta)$ approximately is not an entirely absurd one. Ordinarily n is at least 25 and may be as much as 100 or even larger. If two different solutions $g_1(\zeta)$ and $g_2(\zeta)$ for equation (9) have the same moments up through order 25, these solutions can differ appreciably only if one of them is wavy or jagged. It seems reasonable to assume that the desired solution $g(\zeta)$ shall be "smooth" in some sense. Any smooth solution to (9) should differ very little from any other and thus should provide a useful approximation to the true $g(\zeta)$.

If some mathematical definition of "smoothness" is specified, we can ask which of all possible solutions to (9) is the smoothest.

This is a problem in the calculus of variations. A formula for the smoothest $g(\zeta)$ can sometimes be written down explicitly. In the simplest case (Lord [11]), the smoothest $g(\zeta)$ turns out to be a weighted average of the $h(x|\zeta)$, $x = 0,1,\ldots,n$:

$$\hat{g}(\zeta) = \sum_{x=0}^{n} w_x \, h(x|\zeta \; . \qquad (10)$$

The weights w_x are uniquely determined by the given values of $\phi(x)$.

Such an approach, I think, provides a good approximate solution for the mathematical problem. However, the problem encountered in practice is the statistical problem of estimating $g(\zeta)$ from $f(x)$. What about the obvious procedure of estimating the true-score distribution by finding the smoothest $\hat{g}(\zeta)$ satisfying the equation

$$f(x) = \int_0^1 h(x|\zeta) \, \hat{g}(\zeta) \, d\zeta \; ? \qquad (11)$$

To illustrate the problem, some real data are shown in Table 1 for a conveniently short 15-item test. Although the observed-score distribution apprears reasonably smooth to the eye, the $\hat{g}(\zeta)$ resulting from the suggested procedure is completely useless, as shown in Figure 1.

This sort of result is quite typical. The reason is that the right side of equation (11) represents an operation on $\hat{g}(\zeta)$ that tends to smooth any local irregularities that might be present in $\hat{g}(\zeta)$. The result is that $f(x)$ will be more smooth, in some sense, than $\hat{g}(\zeta)$. The converse must also hold true. The $\hat{g}(\zeta)$ satisfying (11) will tend to be more jagged than the $f(x)$ on the left. When we start with an $f(x)$ containing the usual sample irregularities, these irregularities appear greatly magnified in the $\hat{g}(\zeta)$ obtained from (11). Thus the results shown in Figure 1 are typical.

Another simple procedure would be to assume some special form for $\hat{g}(\zeta)$, to define a "fitted distribution" $\hat{\phi}(x)$ by the equation

$$\hat{\phi}(x) \equiv \int_0^1 h(x|\zeta) \, \hat{g}(\zeta) \, d\zeta \; , \qquad (12)$$

Table 1. Actual observed-score distribution for 3,135 examinees on a 15-item test

x	Nf(x)
15	163
14	324
13	349
12	363
11	352
10	299
9	276
8	236
7	201
6	189
5	126
4	110
3	88
2	40
1	16
0	3

and then to choose the parameters of $\hat{g}(\zeta)$ so as to minimize the chi-square between $\hat{\phi}_x$ and the observed distribution f(x). For convenience one might take $\hat{g}(\zeta)$ to be the natural conjugate Bayes distribution. In the binomial case, then, $\hat{g}(\zeta)$ would be a two-parameter beta distribution. Such an ad hoc procedure might be quite satisfactory for small groups of examinees, but it is utterly inadequate with groups of the size I have studied, which range from N = 700 to n = 300,000. Even for N = 1,000, it was usually found necessary that $\hat{g}(\zeta)$ have at least three or four parameters in order to fit the observed f(x) satisfactorily. I have had partial success using four-parameter beta distributions, but in general this was definitely not adequate, even for N = 1,000.

Clearly, what is needed is to require that $\hat{g}(\zeta)$ be reasonably smooth in some sense, while at the same time providing a good approximate solution to (11). There are many ways that such a procedure might be implemented. Here, I will mention only

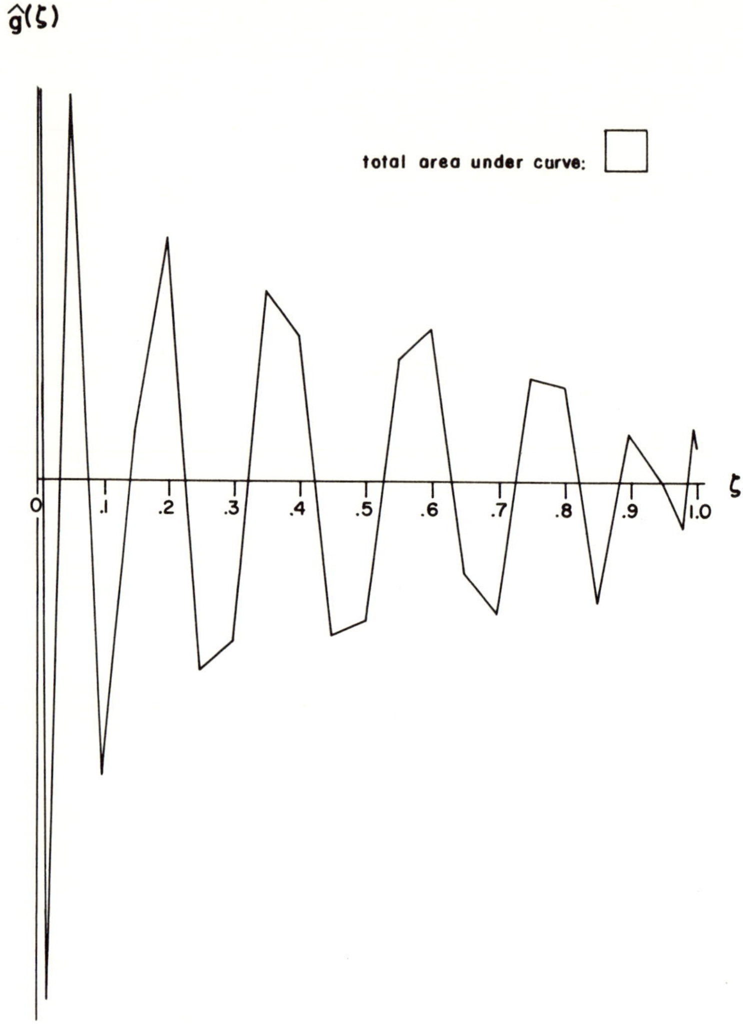

Fig. 1. "Smoothest" $\hat{g}(\zeta)$ giving exact fit to observed-score distribution in sample of 3,135 examinees (two dozen plotted points have simply been connected by straight lines).

the case where an equation similar to (10) is assumed to provide an adequate approximation to $\hat{g}(\zeta)$. Clearly, the w_x are the parameters of the true-score distribution in this case. If these are estimated directly from f(x), the result will be exactly the same as obtained by solving equation (11). The procedure successfully used (Lord [11]) to avoid this difficulty has been to group the observed-score distribution f(x) into class intervals. Summing both sides of equation (12) gives

$$\sum_{x:u} \hat{\phi}(x) \equiv \int_0^1 \hat{g}(\zeta) \sum_{x:u} h(x|\zeta)\, d\zeta \ , \qquad (13)$$

where $\sum_{x:u}$ denotes summation over those values of x grouped together (by the statistician) into class interval u. The "smoothest: solution of the equation now has the form (compare eq. 10)

$$\hat{g}(\zeta) = \sum_{u=1}^{U} w_u \sum_{x:u} h(x|\zeta) \ , \qquad (14)$$

where the first summation is over all class intervals.

If the w_u were estimated directly from the grouped distribution of x, an exact fit to the grouped distribution would be obtained. An even better procedure, however, is to estimate the parameters w_u from the ungrouped f(x). The writer has done this by maximum likelihood. In order to insure that $\hat{g}(\zeta)$ shall not have negative values within the range $0 \leq \zeta \leq 1$, it has been found useful to require that all the w_u be non-negative. This restriction seems to produce smoother $g(\zeta)$, without great damage to the ability of the procedure to fit data. This fit is measured by the (modified) chi-square

$$\dot{\chi}^2 = \sum_{x=0}^{n} \frac{[\hat{\phi}(x) - f(x)]^2}{f(x)} \ . \qquad (15)$$

This procedure has been tried now on about 60 different observed-score distributions (see Lord [11]) with great success. One of the poorest results is shown in Figure 2, which is based on data for about 20,000 examinees. (Here, an obvious modification of the dashed line would improve the fit slightly near the mode, but this would reduce the chi-square only slightly; the main contribution to chi-square is at x = 17, where there is no plausible way to modify the fitted distribution without introducing a bump.)

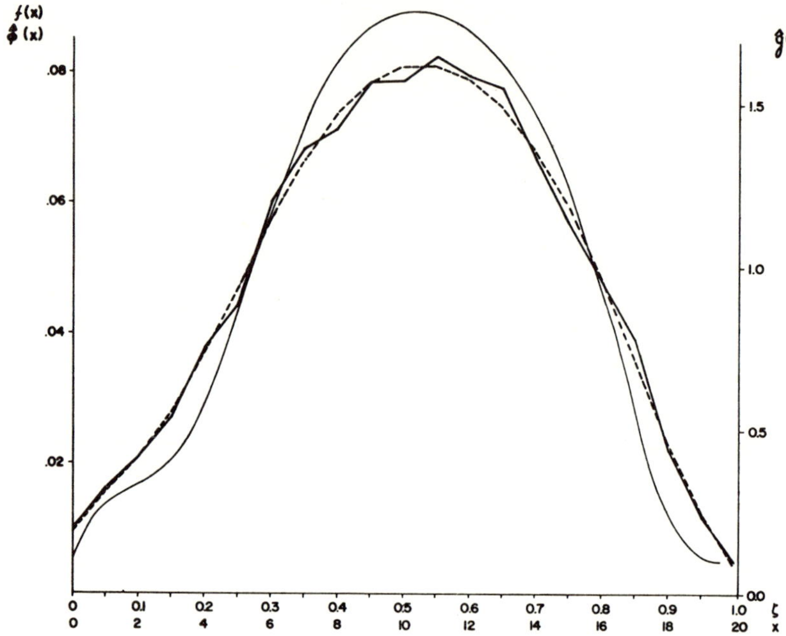

Fig. 2. Observed-score distribution (irregular polygon), fitted observed-score distribution (dotted polygon), and estimated true-score distribution for the worst-fitted set of data in Sample A.

Another illustrative set of data is shown in Figure 3.

It is useful and encouraging to find that reasonably smooth $g(\zeta)$ can be found to fit most sets of data, even with N as large as 100,000 and more. There is nothing in this, however, to give us much confidence in the assumption that $h(x|\zeta)$ is binomial or compound binomial. A much more stringent check on this assumption can be obtained by considering bivariate distributions of observed scores. Consider two tests x and y that are assumed to "measure the same psychological trait," by which we mean that the true scores ζ and η, respectively, are strictly monotonically related. A reasonable model, analogous to (5), is that

$$\phi(x,y) = \int_0^1 g(\zeta)\ h_1(x|\zeta)\ h_2(y|\eta(\zeta))\ d\zeta\ , \tag{16}$$

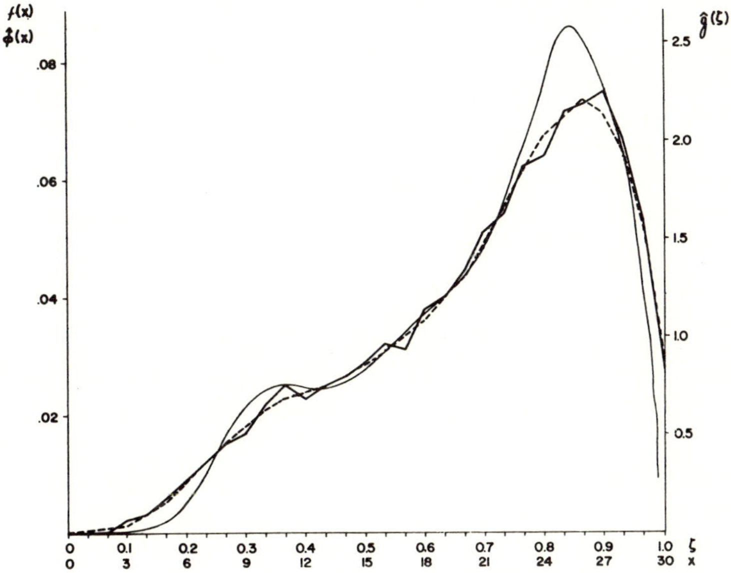

Fig. 3. Observed-score distribution (irregular polygon), fitted observed-score distribution (dotted polygon), and estimated true-score distribution for the worst-fitted set of data in Sample B.

where $\phi(x,y)$ is the bivariate distribution of observed scores in the population, h_1 and h_2 are binomial or compound binomial distribution, and $\eta = \eta(\zeta)$ is the monotonic function relating the two true scores.

Since η and ζ are the same except for the monotonic transformation $\eta(\zeta)$, we have that

$$\int_0^{\zeta} g_1(\zeta)\, d\zeta = \int_0^{\eta} g_2(\eta)\, d\eta \ , \tag{17}$$

where $g_1(\zeta)$ and $g_2(\eta)$ are true-score distributions for tests x and y, respectively. Since $g_1(\zeta)$ and $g_2(\eta)$ can be independently estimated, each from its own observed-score distributions ($f_1(x)$ and $f_2(y)$, say), equation (17) can be used to estimate the monotonic transformation $\eta(\zeta)$. In practice, this is done numerically by inverse interpolation. When $\eta(\zeta)$ has been estimated numerically, then $\phi(x,y)$ can be estimated from equation (16) by numerical integration. Thus the bivariate distribution $\phi(x,y)$ of the observed scores on two tests that measure the same psychological trait can be estimated, under this model, just from the two univariate distributions of the observed scores. The $\hat{\phi}(x,y)$ so obtained from equation (16) can be checked against the actual observed bivariate distribution of x and y to obtain a much more stringent check on the usefulness of the model. Such checks have been carried out for quite a few bivariate distributions with N = 1,000 to 3,000 cases (see Lord [10]). The chi-square between the actual and predicted bivariate distribution has usually been statistically nonsignificant.

The ability, represented by equation (16), to predict bivariate distributions from univariate distributions can be put to practical use. For example, one can administer a full-length form (y) and a short form (x) of a test to a conveniently available sample of examinees and estimate $\eta(\zeta)$ from these data; then administer the short form to a representative national sample of examinees and estimate $g(\zeta)$ from these data; then (under the plausible assumption that $\eta(\zeta)$ is the same in both groups) use (16) to estimate $\phi(x,y)$ for the national group. Now, finally the estimated marginal distribution of y obtained from the estimated $\phi(x,y)$ provides estimated national norms for the full-length form of the test.

Equation (16) is also useful for estimating the effects of using fallible scores for the selection of examinees. It can be used for many other purposes also.

Although I am concerned mainly with binomial error models in mental testing, it seems to me that the model represented by equation (5) plays an important role in many sciences. Surely there will be much productive work in the near future on the statistical inference problems, as distinct from the mathematical problems, raised by equation (5).

REFERENCES

[1] Bakushinski, A. B. 1965. About one numerical method for the solution of Fredholm integral equations of the first kind. U.S.S.R. Comput. Math. and Math. Phys. 5:216-23.

[2] Bullard, E. C., and Cooper, R. I. B. 1948. The determination of the masses necessary to produce a given gravitational field. Proc. Roy. Soc. Ser. A 194:332-5.

[3] Deely, J. J., and Kruse, R. L. 1968. Construction of sequences estimating the mixing distribution. Ann. Math. Statist. 39:286-8.

[4] Eddington, A. S. 1913. On a formula for correcting statistics for the effects of a known probable error of observation. Roy. Astronom. Soc. Monthly Notices 73:359-60.

[5] Harrington, R. F. 1968. Field Computation by Moment Methods. Macmillan Co., New York.

[6] Hudak, J. I. 1966. Regularization of solutions of integral equations of the first kind. (Russian) Z. Vycisl. Mat. i Mat. Fiz. 6:766-9.

[7] Ivanov, V. V., and Kudrinskii, V. J. 1967. Approximate solution of linear operator equations in Hilbert space by the method of least squares. (Russian) Z. Vycisl. Mat. i Mat. Fiz. 7:475-96.

[8] Kreisel, G. 1949. Some remarks on integral equations with kernels: $L(\zeta_1 - x_1, \ldots, \zeta_n - x_n; \alpha)$. Proc. Roy. Soc. Ser. A 197:160-4

[9] Kurth, R. 1965. On Eddington's solution of the convolution integral equation. Rend. Circ. Mat. Palermo. 14:76-84.

[10] Lord, F. M. 1965. A strong true-score theory, with applications. Psychometrika 30:239-70.

[11] ———. 1969. Estimating true-score distributions in psychological testing (An empirical Bayes estimation problem). Psychometrika, in press.

[12] Maiorov, L. V. 1965. The correctness of an inverse problem. (Russian) Ž. Vyčisl. Mat. i Mat. Fiz. 5:363-5.

[13] Maritz, J. S. 1966. Smooth empirical Bayes estimation for one-parameter discrete distributions. Biometrika 53:417-29.

[14] Petrov, A. P. 1967. Estimates of linear functionals for the solution of some inverse problems. (Russian) Z Vycisl. Mat. i Mat. Fiz. 7:648-54.

[15] Phillips, D. L. 1962. A technique for the numerical solution of certain integral equations of the first kind. J. Assoc. Comput. Mach. 9:84-97.

[16] Pollard, H. 1953. Distribution functions containing a Gaussian factor. Proc. Amer. Math. Soc. 4:578-82.

[17] Provencher, S. W. 1967. Numerical solution of linear integral equations of the first kind. Calculation of molecular weight distributions from sedimentation equilibrium data. J. Chem. Phys. 46:3229-36.

[18] Ribiere, G. 1967. Regularization of operators. (French) Rev. Française Informat. Recherche Opérationnelle 1:57-79.

[19] Robbins, H. 1964. The empirical Bayes approach to statistical decision problems. Ann. Math. Statist. 35:1-20.

[20] Suleimanov, R. V. 1967. Solution by the method of successive approximations of the Fredholm integral equation of the first kind, regularized in L_2. (Russian) Ural. Gos. Univ. Mat. Zap. 4:101-4.

[21] Tikhonov, A. N. 1963a. On the solution of incorrectly put problems and the regularization method. (Russian) Outlines Joint Sympos. Partial Differential Equations (Novosibirsk, 1963). Pp. 261-5. Acad. Sci. U.S.S.R. Siberian Branch, Moscow.

[22] ———. 1963b. Solution of incorrectly formulated problems and the regularization method. Soviet Math. Dokl. 4:1035-8.

[23] ———. 1963c. Regularization of incorrectly posed problems. Soviet Math. Dokl. 4:1624-7.

[24] ———. 1964. Solution of nonlinear integral equations of the first kind. Soviet Math. Dokl. 5:835-8.

[25] ———. 1965. On methods for solving incorrect problems encountered in data processing. (Russian) Proc. Conf. Math. Methods of Solution of Problems in Nuclear Phys. Pp. 5-10. Ob"edin. Inst. Jadernyh Issled., Dubna.

[26] Trumpler, R. J., and Weaver, H. F. 1953. Statistical Astronomy. University of California Press, Berkeley.

[27] Turcin, V. F. 1967. The solution of Fredholm's equation of the first kind in the statistical ensemble of smooth functions. (Russian) Z. Vycisl. Mat. i Mat. Fiz. 7:1270-84.

[28] Twomey, S. 1963. On the numerical solution of Fredholm integral equations of the first kind by the inversion of the linear system produced by quadrature. J. Assoc. Comput. Mach. 10:97-101.

[29] ———. 1965. The application of numerical filtering to the solution of integral equations encountered in indirect sensing measurements. J. Franklin Inst. 279:95-109.

[30] Wolfowitz, J. 1957. The minimum distance method. Ann. Math. Statist. 28:75-88.

APPLICATIONS OF A POISSON DISTRIBUTION WITH MULTIPLICATIVE PARAMETER TO INFANT MORTALITY II

JON STENE
Institute of Statistics
University of Copenhagen

SUMMARY

A Poisson distribution with multiple parameters is tested against material from Denmark on infant mortality of different causes during the period 1931-60. It is found that, although this model gives a satisfactory fit to the infant mortality of each cause separately and to the total infant mortality in the period, a combined test on the whole material leads to the rejection of the model. A theoretical reason is given.

1. INTRODUCTION

Let n_{tij} be the number of children born in an area in year t and in group (i,j), where t = 1,...,T, i = 1 denotes male, i = 2, female, j = 1, legitimate birth, and j = 2, illegitimate birth. Further, let x_{tij} denote the number of the corresponding n_{tij} children who have died from cause A in their first year. It was shown in a previous paper (Stene [4]) that the infant mortality in Denmark 1931-60 by noninfectious diseases (except pneumonia) could be described by the model

$$p\{X_{tij}=x_{tij}|\xi_{tj},\varepsilon_{ij}\} = \frac{(n_{tij}\xi_{tj}\varepsilon_{ij})^{x_{tij}}}{x_{tij}!} e^{-n_{tij}\xi_{tj}\varepsilon_{ij}}, \quad (1)$$

where X_{tij} is a stochastic variable and ξ_{tj} is a parameter for the general infant mortality due to cause A in the year t for legitimate births if j = 1 and for illegitimate births if j = 2. ξ_{tj} may be considered as a measure of medical and social conditions in year t for this legitimacy group. ε_{1j} and ε_{2j} are the specific infant mortalities for males and females, respectively, in the actual legitimacy group and are by hypothesis independent of t, but may vary from cause to cause.

It was also shown that, since the sex ratio for each legitimacy group could be considered approximately constant during the period, the conditional probability distribution of the number of boys X_{t1j} born in year t and in legitimacy group j, who die from cause A during their first year (given the total infant mortality among the children born in year t in legitimacy group j by this cause, is binomial

$$p\{X_{t1j}=x_{t1j}|x_{t.j},\tau_j\} = \binom{x_{t.j}}{x_{t1j}}\tau_j^{x_{t1j}}(1-\tau_j)^{x_{t2j}}) \quad (2)$$

where the parameter τ_j is independent of t and $x_{t.j} = x_{t1j}+x_{t2j}$.

The aim of this paper is to carry out a more thorough analysis of the infant mortality material from Denmark in order to get an idea of the applicability of the model (1) to this type of material.

2. ANALYSIS

We want to test the hypothesis that τ_j remains constant during the period 1931-60 for a given cause and legitimacy group.

This test is based on the χ^2-statistic Z:

$$Z = \sum_{t=1}^{T} \frac{\left(x_{t1j} - x_{t.j}\frac{x_{.1j}}{x_{..j}}\right)^2}{x_{t.j}\frac{x_{.1j}}{x_{..j}}\left(1 - \frac{x_{.1j}}{x_{..j}}\right)} ; \tag{3}$$

which is approximately χ^2-distributed with (T-1) df. In (3) we have introduced

$$x_{.1j} = \sum_{t=1}^{T} x_{t1j}, \quad x_{..j} = \sum_{t=1}^{T} x_{t.j} \quad .$$

By (3) we test the hypothesis that the conditional distribution of X_{t1j}, given $X_{t.j} = X_{t1j}+X_{t2j} = x_{t.j}$ is binomial distributed with the parameter independent of t. Rasch [3] has proved that if this hypothesis holds, then the distribution must be of the form (1). This result is also a consequence of theorem 1 in Chatterji [1].

The material is taken from Matthiessen [2] and given in tables 1-7. Tables 1 and 2 present the material for different noninfectious causes except pneumonia for legitimate and illegitimate births, respectively. Tables 3 and 4 give the material for different infectious diseases for the two legitimacy groups. For the classifications and their variation see Matthiessen [2]. For the years 1951-1960 noninfectious diseases are not considered separately because the classification had changed. In tables 5 and 6 the infant mortality due to pneumonia is given together with the marginals from the preceding tables. Table 7 presents the total infant mortality for the two legitimacy groups.

In table 8 the values of the χ^2 statistics for the material in tables 1 and 2 are given. In table 9 the corresponding values for the material in tables 3 and 4 are presented. For influenza and whooping cough in illegitimate cases, the numbers are so small that the approximation to the χ^2-distribution is invalid. For several of the other causes the material is reduced in order to avoid this same problem.

By inspecting the tables we see that none of the χ^2-values exceeds the upper 5%-point in the corresponding distribution. Therefore we conclude that for each cause and legitimacy group the model (1) gives a satisfactory fit.

Then we want to test the hypothesis that the model (1) simultaneously gives a satisfactory fit, but the ε_{ij}'s and the ξ_{tj}'s may vary from cause to cause.

In order to test this hypothesis we add the values in the tables 8 and 9 and for pneumonia from table 10. As a result, we get the χ^2-statistic $Z = 473.6141$ with 424 df. This result is significant at 5% level.

We must conclude, therefore, that the model does not give a satisfactory fit for the whole materiale simultaneously, but the marginals show a good fit. This result represents a nice practical example of a theoretical result derived by Walsh [5] that the Poisson distribution is a limit for dependent binomial events with unequal probabilities when the sum of the probabilities over all events is small compared with the number of observations and when each event is independent of a large number of the other events. The causes which lead to a rejection are mainly the infectious ones where some dependence is present.

ACKNOWLEDGEMENTS

The author wishes to thank his wife, Dipl.ing. Eeva Stene, for numerical work and Mrs. Inge Hallgren for typing the manuscript.

REFERENCES

[1] Chatterji, S. D. 1963. Some elementary characterizations of the Poisson-distribution. Am. Math. Month. 70:958-64.

[2] Matthiessen, P. C. 1965. Infant mortality in Denmark 1931-60. Copenhagen. The Statistical Department.

[3] Rasch, G. 1968. En reanalyse af danske og svenske forsøg over virkningen af hastighedsbegraensninger på trafikulykker. Mimeographed technical report. Inst. of Stat. Copenhagen.

[4] Stene, J. 1969. An application of a Poisson distribution with multiplicative parameter to infant mortality. Contributed paper at the 37th Session of the ISI.

[5] Walsh, J. E. 1955. The Poisson distribution as a limit for dependent binomial events with unequal probabilities. J. Op. Res. Soc. Am. 3:198-209.

Table 1. Infant Mortality for Noninfectious Diseases (excl. pneumonia), Legitimate Birth

	Legitimate Birth																	
	Congenital Malformations			Prematurity			Congenital Debility			Injuries at Birth			Other Infants' Diseases			Other Noninfectious Diseases		
	M	F	M+F	M	F	M+F	M	F	M+F	M	F	M+F	M	F	M+F	M	F	M+F
1931	194	156	350	375	278	653	172	118	290	103	45	148	45	33	78	498	302	800
1932	179	152	331	376	274	650	179	122	301	85	67	152	49	42	91	323	255	578
1933	197	168	365	325	262	587	163	110	273	102	43	145	39	23	62	336	205	541
1934	194	159	353	363	273	636	202	117	319	119	61	180	68	39	107	335	230	565
1935	224	154	378	413	307	720	197	110	307	100	52	152	62	44	106	316	221	537
1936	229	170	399	386	284	670	211	138	349	101	62	163	54	34	88	279	197	476
1937	244	198	442	412	300	712	187	120	307	109	64	173	61	34	95	313	178	491
1938	221	173	394	345	274	619	209	158	367	116	74	190	52	37	89	249	163	412
1939	190	196	386	327	256	583	230	147	377	115	81	196	48	41	89	243	152	395
1940	222	175	397	375	249	624	201	164	365	109	56	165	51	34	85	190	165	355
1941	234	186	420	402	252	654	151	95	246	118	69	187	58	43	101	224	195	419
1942	213	192	405	464	281	745	162	115	277	94	51	145	94	54	148	245	168	413
1943	239	181	420	449	299	748	163	118	281	104	44	148	92	60	152	219	158	377
1944	265	187	452	545	392	937	179	115	294	105	50	155	90	56	146	242	154	396
1945	304	217	521	525	389	914	179	102	281	106	66	172	84	61	145	229	180	409
1946	270	198	468	566	464	1030	103	57	160	149	88	237	96	47	143	257	173	430
1947	239	168	407	534	363	897	59	52	111	134	78	212	123	64	187	237	187	424
1948	205	155	360	453	338	791	53	25	78	117	63	180	83	46	129	212	142	354
1949	198	125	323	391	328	719	49	34	83	122	57	179	92	57	149	150	101	251
1950	194	138	332	373	273	646	61	29	90	102	54	156	76	62	138	160	103	262
Total	4455	3448	7903	8399	6136	14535	3110	2046	5156	2210	1225	3435	1417	911	2328	5257	3629	8886

Table 2. Infant Mortality for Noninfectious Diseases (excl. pneumonia), Illegitimate Birth

	Congenital Malformations			Prematurity			Congenital Debility			Injuries at Birth			Other inf. Diseases			Other Noninfectious Diseases		
	M	F	M+F	M	F	M+F	M	F	M+F	M	F	M+F	M	F	M+F	M	F	M+F
1931	25	18	43	80	77	157	34	18	52	19	2	21	7	7	14	93	56	149
1932	16	18	34	87	81	168	34	25	59	7	5	12	5	6	11	62	47	109
1933	16	13	29	83	67	150	32	26	58	12	5	17	4	0	4	52	42	94
1934	15	11	26	72	65	137	28	15	43	12	9	21	3	5	8	49	39	88
1935	26	26	52	87	58	145	26	15	41	16	9	25	4	2	6	48	33	81
1936	13	21	34	100	70	170	20	14	34	17	13	30	4	3	7	45	45	90
1937	20	22	42	95	77	172	15	10	25	11	9	20	4	2	6	33	22	55
1938	17	23	40	73	73	146	40	21	61	11	5	16	4	3	7	42	37	79
1939	17	19	36	70	47	117	35	25	60	10	8	18	4	7	11	52	30	82
1940	28	17	45	82	55	137	30	13	43	10	8	18	4	4	8	29	27	56
1941	21	21	42	78	36	114	33	10	43	11	2	13	4	0	4	42	26	68
1942	30	15	45	95	67	162	28	17	45	8	10	18	10	5	15	45	36	81
1943	27	12	39	101	76	177	23	21	44	12	4	16	10	7	17	36	15	51
1944	28	20	48	109	104	213	34	24	58	10	8	18	9	8	17	30	39	69
1945	34	27	61	140	107	247	35	26	61	17	11	28	8	7	15	36	34	70
1946	20	24	44	109	84	193	21	14	35	16	5	21	6	7	13	55	31	86
1947	38	13	51	100	79	179	7	5	12	13	7	20	6	6	12	46	32	78
1948	14	11	25	65	48	113	9	5	14	6	7	13	5	4	9	26	30	56
1949	17	10	27	75	56	131	6	0	6	9	6	15	4	3	7	23	10	33
1950	15	17	32	85	53	138	5	2	7	10	6	16	10	6	16	14	7	21
Total	437	358	795	1786	1380	3166	495	306	801	237	139	376	115	92	207	858	638	1496

Table 3. Infectious Diseases, Legitimate Birth

	Influenza			Cholerine			Whooping Cough			Other Infectious Diseases		
	M	F	M+F	M	F	M+F	M	F	M+F	M	F	M+F
1931	48	66	114	236	141	377	97	85	182	118	87	205
1932	41	54	95	237	153	390	110	143	253	124	78	202
1933	61	37	98	181	144	325	46	47	93	114	76	190
1934	14	13	27	184	139	323	36	32	68	111	80	191
1935	35	32	67	199	132	331	93	83	176	176	148	324
1936	45	29	74	191	123	314	107	81	188	124	104	228
1937	48	34	82	188	132	320	27	33	60	124	118	242
1938	25	18	43	150	116	266	21	26	47	140	94	234
1939	33	17	50	143	79	222	62	68	130	107	75	182
1940	26	18	44	108	66	174	32	42	74	84	51	135
1941	35	25	60	106	69	175	66	50	116	108	79	187
1942	10	6	16	95	67	162	37	43	80	97	61	158
1943	5	7	12	100	72	172	45	45	90	106	67	173
1944	16	13	29	159	115	274	37	37	74	143	107	250
1945	8	5	13	210	149	359	45	48	93	146	112	258
1946	5	10	15	129	83	212	49	63	112	107	85	192
1947	5	8	13	78	38	116	10	18	28	97	69	166
1948	2	2	4	59	29	88	41	40	81	77	51	128
1949	12	8	20	46	24	70	55	51	106	58	43	101
1950	4	6	10	32	23	55	4	5	9	43	35	78
1951	0	2	2	36	22	58	7	19	26	37	25	62
1952	2	1	3	35	29	64	17	19	36	34	18	52
1953	6	5	11	21	26	47	10	10	20	33	22	55
1954	8	2	10	41	23	64	5	5	10	26	13	39
1955	0	1	1	28	22	50	2	1	3	12	23	35
1956	1	1	2	28	8	36	9	7	16	20	11	31
1957	5	1	6	29	16	45	8	6	14	22	18	40
1958	5	2	7	15	15	30	1	0	1	15	14	29
1959	6	5	11	17	18	35	2	4	6	17	16	33
1960	2	1	3	12	15	27	4	3	7	15	15	30
	513	429	942	3093	2088	5181	1085	1114	2199	2435	1795	4230

Table 4. Infectious Diseases, Illegitimate Birth

	Influenza			Cholerine			Whooping Cough			Other Infectious Diseases		
	M	F	M+F	M	F	M+F	M	F	M+F	M	F	M+F
1931	4	7	11	26	27	53	11	17	28	21	14	35
1932	9	7	16	53	23	76	10	15	25	13	11	24
1933	3	8	11	39	14	53	6	8	14	15	14	29
1934	4	1	5	35	13	48	3	4	7	12	7	19
1935	5	7	12	31	19	50	11	3	14	17	13	30
1936	4	1	5	23	14	37	15	9	24	20	7	27
1937	3	1	4	13	8	21	3	1	4	13	12	25
1938	6	1	7	20	10	30	4	1	5	16	7	23
1939	1	1	2	16	10	26	4	4	8	9	13	22
1940	2	2	4	11	10	21	1	5	6	11	9	20
1941	2	3	5	17	8	25	7	5	12	10	4	14
1942	1	0	1	18	5	23	0	2	2	14	10	24
1943	1	2	3	27	18	45	5	6	11	13	16	29
1944	0	1	1	43	25	68	4	5	9	28	20	48
1945	0	1	1	68	38	106	5	1	6	29	15	44
1946	1	0	1	53	41	94	6	7	13	13	15	28
1947	3	0	3	25	13	38	0	0	0	12	4	16
1948	1	0	1	4	8	12	3	1	4	7	5	12
1949	1	0	1	9	5	14	2	3	5	8	2	10
1950	0	0	0	7	4	11	0	2	2	3	5	8
1951	1	0	1	3	2	5	0	1	1	3	3	6
1952	1	0	1	3	1	4	1	1	2	1	2	3
1953	0	0	0	4	1	5	0	0	0	6	0	6
1954	0	0	0	3	3	6	1	0	1	2	1	3
1955	0	0	0	2	1	3	1	0	1	3	2	5
1956	0	0	0	5	2	7	0	0	0	2	2	4
1957	0	1	1	3	3	6	1	3	4	0	0	0
1958	0	0	0	3	2	5	0	0	0	1	3	4
1959	1	1	2	2	2	4	0	0	0	2	2	4
1960	0	0	0	3	0	3	2	0	2	2	1	3
	54	45	99	569	330	899	106	104	210	306	219	525

Table 5. Legitimate Birth, Marginals and Pneumonia

	Noninfectious (excl. pneu.)			Pneumonia			Infectious Diseases		
	M	F	M+F	M	F	M+F	M	F	M+F
1931	1387	932	2319	758	538	1296	499	379	878
1932	1191	912	2103	549	413	962	512	428	940
1933	1162	811	1973	566	419	985	402	304	706
1934	1281	879	2160	544	376	920	345	264	609
1935	1312	888	2200	565	419	984	503	395	898
1936	1260	885	2145	552	404	956	467	337	804
1937	1326	894	2220	593	443	1036	387	317	704
1938	1192	879	2071	512	340	852	336	254	590
1939	1153	873	2026	499	353	852	345	239	584
1940	1148	843	1991	385	288	673	250	177	427
1941	1187	840	2027	523	390	913	315	223	538
1942	1272	861	2133	398	284	682	239	177	416
1943	1266	860	2126	402	290	692	256	191	447
1944	1426	954	2380	358	321	679	355	272	627
1945	1427	1015	2442	402	287	689	409	314	723
1946	1441	1027	2468	458	317	775	290	241	531
1947	1326	912	2238	373	291	664	190	133	323
1948	1123	769	1892	289	196	485	179	122	301
1949	1002	702	1704	257	203	460	171	126	297
1950	966	659	1625	226	160	386	83	69	152
1951	926	645	1571	158	117	275	80	68	148
1952	951	638	1589	124	114	238	88	67	155
1953	967	622	1589	99	93	192	70	63	133
1954	906	606	1512	117	98	215	80	43	123
1955	874	644	1518	72	68	140	42	47	89
1956	910	606	1516	77	47	124	58	27	85
1957	809	522	1331	78	45	123	64	41	105
1958	764	515	1279	76	49	125	36	31	67
1959	760	516	1276	64	46	110	42	43	85
1960	743	551	1294	45	50	95	33	34	67
	33458	23260	56718	10119	7459	17578	7126	5426	12552

Table 6. Illegitimate Birth, Marginals and Pneumonia

	Noninfectious Diseases			Pneumonia			Infectious Diseases		
	M	F	M+F	M	F	M+F	M	F	M+F
1931	258	178	436	114	62	176	62	65	127
1932	211	182	393	77	57	134	85	56	141
1933	199	153	352	73	50	123	63	44	107
1934	179	144	323	66	37	103	54	25	79
1935	207	143	350	61	35	96	64	42	106
1936	199	166	365	69	41	110	62	31	93
1937	178	142	320	68	53	121	32	22	54
1938	187	162	349	47	48	95	46	19	65
1939	188	136	324	49	52	101	30	28	58
1940	183	124	307	30	38	68	25	26	51
1941	189	95	284	59	42	101	36	20	56
1942	216	150	366	49	41	90	33	17	50
1943	209	135	344	46	37	83	46	42	88
1944	220	203	423	53	34	87	75	51	126
1945	270	212	482	51	46	97	102	55	157
1946	227	165	392	58	45	103	73	63	136
1947	210	142	352	32	43	75	40	17	57
1948	125	105	230	34	26	60	15	14	29
1949	134	85	219	22	23	45	20	10	30
1950	139	91	230	18	13	31	10	11	21
1951	99	81	180	15	7	22	7	6	13
1952	104	99	203	19	9	28	6	4	10
1953	110	79	189	7	9	16	10	1	11
1954	105	66	171	10	10	20	6	4	10
1955	101	71	172	2	4	6	6	3	9
1956	97	71	168	3	7	10	7	4	11
1957	98	78	176	5	7	12	4	7	11
1958	104	78	182	5	8	13	4	5	9
1959	102	66	168	5	6	11	5	5	10
1960	99	64	163	6	3	9	7	1	8
	4947	3666	8613	1153	893	2046	1035	698	1733

Table 7.

	Legitimate Total Infant Mortality			Illegitimate Total Infant Mortality		
	M	F	M+F	M	F	M+F
1931	2644	1849	4493	434	305	739
1932	2252	1753	4005	373	295	668
1933	2130	1534	3664	335	247	582
1934	2170	1519	3689	299	206	505
1935	2380	1702	4082	332	220	552
1936	2279	1626	3905	330	238	568
1937	2306	1654	3960	278	217	495
1938	2040	1473	3513	280	229	509
1939	1997	1465	3462	267	216	483
1940	1783	1308	3091	238	188	426
1941	2025	1453	3478	284	157	441
1942	1909	1322	3231	298	208	506
1943	1924	1341	3265	301	214	515
1944	2139	1547	3686	348	288	636
1945	2238	1616	3854	423	313	736
1946	2189	1585	3774	358	273	631
1947	1889	1336	3225	282	202	484
1948	1591	1087	2678	174	145	319
1949	1430	1031	2461	176	118	294
1950	1275	888	2163	167	115	282
1951	1164	830	1994	121	94	215
1952	1163	819	1982	129	112	241
1953	1136	778	1914	127	89	216
1954	1103	747	1850	121	80	201
1955	988	759	1747	109	78	187
1956	1045	680	1725	107	82	189
1957	951	608	1559	107	92	199
1958	876	595	1471	113	91	204
1959	866	605	1471	112	77	189
1960	821	635	1456	112	68	180
	50703	36145	86848	7135	5257	12392

Table 8. χ^2-values for Different Noninfectious Diseases (excl. pneumonia)

	Legitimate	df	Illegitimate	df	Sum	df
Congenital malformations	21.0354	19	28.1120	19	49.1474	38
Prematurity	24.2244	19	20.2307	19	44.4551	38
Congenital debility	19.4724	19	14.6804	19	34.1528	38
Injuries at Birth	18.8551	19	19.7959	19	38.6510	38
Other infants' diseases	13.7995	19	12.2766	19	26.0761	38
Other noninfectious diseases	29.9327	19	24.4907	19	54.4234	38
Sum	127.3195	114	119.5863	114	246.9058	228

Table 9. χ^2-values for the Different Infectious Diseases

	Legitimate	df	Illegitimate	df	Sum	df
Influenza	25.3295*	16	7.2156***	5	32.5451	21
Cholerine	32.3282	29	16.4009*	16	48.7291	45
Whooping Cough	24.4456**	21	8.7857***	5	33.2313	26
Other infectious diseases	25.6516	29	15.7945*	16	41.4461	45
Sum	107.7549	96	48.1967	42	155.9516	138

* For the years 1931-47

** For the years 1931-53

*** For the years 1931-36

Table 10. The Three Main Groups of Causes in 1931-60

	Legitimate	df	Illegitimate	df	Sum	df
Noninfectious	24.7877	29	35.7370	29	60.5247	58
Pneumonia	30.0366	29	40.7201	29	70.7567	58
Infectious	26.3283	29	41.5835	29	67.9118	58
Sum	81.1526	87	118.0406	87	199.1932	174
Total Infant Mortality	28.7317	29	26.0992	29	54.8309	58

PLAUSIBILITY AND SECOND-ORDER ENTROPY, WITH PARTICULAR REFERENCE TO FAMILY COMPOSITION

ADRIENNE W. KEMP
The Queen's University of Belfast

SUMMARY

This paper considers a problem posed by Gridgeman [2] about the surprise value concerning the sex composition of a particular Illinois family. The concept of the plausibility associated with a discrete random count is defined as the reciprocal of Weaver's Surprise Index, and is related to

(a) Fry's expectation of a probability,

(b) Nil'son's k-variance, and

(c) Réyni's second-order entropy.

1. INTRODUCTION

In a recent article entitled "Probability and Sex" Gridgeman [2] has examined the probability statements which can be made concerning an Illinois family consisting of five girls followed by six boys. Throughout his article Gridgeman assumes that each child has an equal, independent chance ($\frac{1}{2}$) of being a boy. Although it is arguable that boys are more likely than girls, and that the sex of subsequent children is not independent of that of preceding children in a family, Gridgeman's assumptions will again be adopted. Primarily this paper is concerned with certain statistical concepts which do not often appear in the literature; the Illinois family is regarded purely as an illustrative example.

Gridgeman rightly comments that "a 5:6 sex-split in 11 siblings has of itself no surprise value, being, in fact, the most likely event". Conversely we could attribute the greatest surprise value to the least likely event, namely a 0:11 sex-split. However, what surprise value do we ascribe to a sex-split of say 2:9?. The probability of this event (on the given assumptions) is approximately 0.054. However, as Weaver [7] has put it, "The mere fact that a probability is low should not of itself lead to incredulity .... One must always, before concluding that a certain event is surprising, examine the probability of the alternatives". In this example the average probability $E(p_i)$ (= $\sum_i p_i^2$) is approximately 0.336, giving, for a 2:9 sex-split, a ratio of the actual to the expected probability of 0.054/0.336 ≑ 1/6. We conclude that, while a sex-split of 2:9 is a somewhat improbable event, it cannot be regarded as a very surprising event.

2. PLAUSIBILITY AND THE SURPRISE INDEX

The concept of surprise was made more precise by Weaver by his definition of a Surprise Index, which he defined as the ratio of the expected to the observed probability. For Weaver, "A Surprise Index of 3 or 5 is surely not large: one of 10 begins to be surprising: one of 1,000 is definitely surprising: one of 1,000,000 or larger is very surprising indeed: one of 10^{12} would presumably qualify as a miracle."

However, as we shall see, there are certain advantages in

taking the reciprocal of Weaver's Surprise Index; this will be called the plausibility and may be defined as $p_i / \sum_i p_i^2$. Since this depends on $\sum_i p_i^2$, the plausibility of an event, unlike its probability, is dependent on the complete specification of the probabilities of the alternative events.

3. TOTAL PLAUSIBILITY, EXPECTED PROBABILITY, AND NIL'SON'S k-VARIANCE

The sum of the plausibilities associated with the possible outcomes of a random experiment, i.e., total plausibility, will not in general be unity. Since $\sum_i p_i = 1$, total plausibility is the reciprocal of the expected probability, the meaning of which has been discussed by Fry [1]. Fry found that the expected probability is least when all events are equally likely.

Weaver noted that "it seems to be difficult indeed to calculate, in general terms, the expectation of the probability for either the binomial or the Poisson case." It has been found, however, for both distributions by Redheffer [5]. For the Poisson distribution with parameter λ,

$$\sum_i p_i^2 = e^{-2\lambda} I_o\,(-\,2\lambda),$$

and for the binomial distribution with parameters n, p, q (= 1 - p)

$$\sum_i p_i^2 = (p-q)^{2n}\ {}_2F_1\ [-n, \tfrac{1}{2}; 1; -4pq/(p-q)^2], \quad (p \neq q),$$

(an alternative formula was given by Pólya-Szegö, see [5]).

Total plausibility is called the k-variance by Nil'son [4], who not only showed that there is an upper bound n which is attained (for a discrete distribution) iff $p_i = 1/n$ $(i = 1,2,\ldots,n)$, but also that there is a lower bound equal to unity which is attained iff some $p_i = 1$; i.e., Nil'son established that the total plausibility is only unity for a degenerate distribution and is otherwise greater than unity. Nil'son computed the k-variance for the normal distribution; he also discussed its relationship to various measures of dependence and correlation. His paper appears in Russian (with English summary) in an Estonian journal; an extended summary is given by Gundy [3].

4. LOG PLAUSIBILITY AND REYNI'S SECOND-ORDER ENTROPY

The entropy of a distribution measures the amount of uncertainty associated with that distribution. The concept of entropy of order α for a finite discrete distribution ρ was introduced by Réyni (see [6]), who defined it as

$$H_\alpha [\rho] = \frac{1}{(1-\alpha)} \log_2 [\sum_i p_i^\alpha / \sum_i p_i].$$

As $\alpha \to 1$ this becomes the familiar Shannon measure of entropy (first-order entropy)

$$\lim_{\alpha\to 1} H_\alpha [\rho] = H_1[\rho] = \sum_i p_i \log_2 p_i / \sum_i p_i,$$

which, however, does not yield simple expressions for the common discrete distributions.

When $\alpha = 2$, we obtain the second-order entropy

$$H_2[\rho] = \log_2 [\sum_i p_i / \sum_i p_i^2],$$

i.e., the logarithm to base 2 of the total plausibility. Hence the second-order entropies for the Poisson and binomial distributions are obtainable from Redheffer's expressions for the expectations of their probabilities.

Now consider a sample from a discrete distribution with frequencies N_i, $i = 1,2,\ldots,$ such that $\sum_i N_i = N$. The probability (likelihood) of this sample is $\prod_i p_i^{N_i}$; i.e., L (the logarithm of the likelihood) is $\sum_i N_i \log_e p_i$. The second-order entropy of the distribution of such samples of size N is

$$\log_2 [1 / (\sum_i p_i^2)^N], \text{ and putting } \sum_i p_i = 1,$$

$$= N \log_2 [1 / \sum_i p_i^2].$$

Hence the logarithm (to base e) of the plausibility of the sample is

$$\log_e [\prod_i p_i^{N_i} / (\sum_i p_i^2)^N]$$

$$= L + NC\, H_2 [\rho],$$

where $C = \log_e 2$ and $H_2 [\rho]$ is the second-order entropy for a single observation.

5. THE SEQUENTIAL ARRANGEMENT OF THE ILLINOIS FAMILY

Returning to the Illinois family, the probability of obtaining the sample FFFFFMMMMMM (each observation being drawn from the distribution with $p_F = \frac{1}{2}$, $p_M = \frac{1}{2}$) is $1/2^{11}$. The expected probability is $[(\frac{1}{2})^2 + (\frac{1}{2})^2]^{11} = 1/2^{11}$, the second-order entropy is 11, and the plausibility of obtaining the given arrangement compared with <u>any</u> other possible arrangement is, not unexpectedly, unity.

However, if we are only interested in the arrangements (FFFFFMMMMMM, MMMMMFFFFFF, MMMMMMFFFFF, FFFFFFMMMMM) as unusual as that of the Illinois family (probability = $4/2^{11}$) compared with <u>all</u> other possible arrangements (probability = $(2^{11} - 4)/2^{11}$), the expected probability is nearly unity, and the second-order entropy is approximately 0.006. The plausibility is

$$\frac{4}{2^{11}} \times \frac{(2^{11})^2}{[4^2 + (2^{11} - 4)^2]} \doteqdot 0.002;$$

this is decidedly low, and we may regard the family as decidedly surprising.

It is important to note that, whereas the probability of an event is unaffected by alterations in the grouping of the possible alternative events, the concept of plausibility (and hence surprise) depends very much on the structure of the set of alternative events. The plausibility (and surprise value) of an event has no meaning except with reference to an individually specified set of alternative events.

Conditional plausibilities, like conditional probabilities, are sometimes more meaningful than unconditional ones. For example, it could be argued that we have already decided that a 5:6 sex-split is not surprising, and hence that we should consider only the probabilities conditional on such a sex-split. The total number of arrangements conditional on a 5:6 sex-split is 924, the conditional probability of an arrangement as unusual as the given one is 4/924, and the conditional alternative probability is 920/924. The expected probability is again close to unity, the second-order entropy is approximately 0.013, and the conditional plausibility is

$$\frac{4}{924} \times \frac{924^2}{[4^2 + 920^2]} \doteqdot 0.004.$$

As we would expect, this plausibility conditional on a 5:6 sex-split is higher than the unconditional plausibility. However, it is still low, and the family remains surprising.

REFERENCES

[1] Fry, T. C. 1964. Probability and Its Engineering Uses. 2nd Ed., D. Van Nostrand, Princeton, New Jersey.

[2] Gridgeman, N. T. 1968. Probability and Sex. American Statistician 22(3), 27.

[3] Gundy, R. F. 1966. Review of Nil'son, A., Some properties of sums of squares of probabilities and their applications in mathematical statistics. Mathematical Reviews 32, 4763.

[4] Nil'son, A. 1965. Some properties of sums of squares of probabilities and their applications in mathematical statistics. (Russian with Estonian and English summaries), Eesti N.S.V. Tead. Akad. Toimetised Fuü. -Mat. Tehn. -tead. Seer. 14, 79-93.

[5] Redheffer, R. M. 1951. A note on the surprise index. Ann. Math. Statist. 22, 128-30.

[6] Réyni, A. 1961. On measures of entropy and information. Proc. 4th Berkeley Symp. on Math. Stats. and Prob. I, 547-61.

[7] Weaver, W. 1948. Probability, rarity, interest, and surprise. Scientific Monthly 67, 390-92.

PROBLEMS OF SPATIAL DISTRIBUTION IN ANIMAL POPULATION ECOLOGY

SYUN'ITI IWAO
Entomological Laboratory
Kyoto University
Kyoto, Japan

SUMMARY

The problems of spatial distribution of animals are reviewed and discussed with particular reference to the statistical methods available. Application of discrete distribution models to animal distributions is useful for determining sampling design and transformation of data, but it does not suggest much about the underlying mechanisms of the observed distribution. Aggregation indices based on the first two moments alone may often be more convenient, because they give information on the nature of distribution without any *a priori* assumption. Particularly, the regression method based on the relation of mean crowding to mean density provides a comprehensive way to describe and analyze the spatial pattern of a species, in which basic patterns generated by various mathematical models are incorporated as special cases. It also gives general methods for determining sample size and transformation of data. Perspectives of several important topics are discussed briefly.

1. INTRODUCTION

The individuals of a species of animal distribute themselves in their habitat with a characteristic pattern, depending both on the inherent properties of the species and the conditions of the habitat. The analysis of spatial distribution is very important in animal population ecology for the following two reasons:

(1) the pattern of spatial distribution affects not only the precision of the estimation of population parameters in sampling surveys, but also the method of analysis of the data; and

(2) the spatial distribution itself is an important structure of the population. The distribution at a given time is determined by complex biological processes that were going on in the population by that time, and it may define or affect the subsequent change of the population. Therefore, the information on the spatial pattern and its change in relation to population density and various environmental variables are fundamental for a full understanding of animal population dynamics.

Theoretically, the distribution of individuals in a space could be most properly treated by the distance methods (cf. [14], [47] - [50], [61], [62], and [18], but in animals it is usually analyzed by using the frequency counts of individuals in habitat units (quadrats). This is because animals live in three dimensional space, and their habitat is not continuous but can be divided into a number of discrete units such as individual plants, leaves, fruits, etc. The scope of this paper, therefore, will be confined to the distribution of animals in habitat units, and the topics concerning the application of mathematical models and the measures of aggregation will be reviewed and discussed in an arbitrary manner.

2. APPLICATION OF MATHEMATICAL DISTRIBUTIONS

2.1 Poisson Distribution and the Variance-Mean Ratio

The simplest assumption about the spatial distribution of individuals in a space would be that a number of organisms are scattered independently and at random. Since an organism usually occupies a negligible portion of a quadrat and the total number of quadrats in the area is large, the Poisson distribution is appropriate for this assumption. In this distribution the expected variance (σ^2) equals the mean (m), and hence the ratio of observed variance (s^2) to the sample mean ($\bar{x}$) has been widely used as a convenient criterion to judge whether or not the observed distribution can be considered random (see p. 17).

It is now well-known that organisms, both animals and plants, are not often distributed randomly even in a seemingly homogeneous habitat. For example, Cole [16], who studied the spatial pattern of several cryptozoic animals under boards, found that only the distributions of spiders regularly corresponded to a Poisson distribution, but all of the other important species were more or less clumped in their distributions.

Although at low mean densities many animals are reported to have a distribution that does not differ significantly from the Poisson, it does not necessarily mean that the animals are distributed randomly under these conditions. It seems mainly due to a statistical deficiency that the frequency distribution of a small number of organisms usually cannot depart noticeably from the Poisson expectation even if it has some aggregation tendency (e.g., Waters and Henson [85]).

2.2 Negative Binomial Distribution and Other Contagious Distributions

Overdispersed or aggregated distributions that are commonly observed in animal populations may arise from various causes: mutual attraction between individuals, such as sexual attraction and gregarious habits; heterogeneity in the habitat in relation to animal's preference and differential survival rates; reproductive behavior, such as collective laying of eggs or care of young and limited ability for dispersal of subsequent stages; and resulting patterns of population growth over generations (e.g., Waters [84], Southwood [71]).

Among various contagious distribution models hitherto proposed, the negative binomial series is proved to be most widely applicable to aggregated biological distributions (e.g., Bliss and Fisher [8]; Bliss [7]; and Waters [83]). It is defined by two parameters, the arithmetic mean m and a positive exponent k, and given by the extension of $(q-p)^{-k}$, where $p = m/k$ and $q = 1+p$. The distribution is flexible in the sense that it converges to the Poisson series when $k \to \infty$, whereas it approaches the logarithmic series when $k \to 0$. Also, it can be derived in several different ways (Anscombe [1]):

(1) True contagion, where the presence of one individual in a unit increases the probability that another individual will enter the same unit (known as Pólya-Eggenberger's distribution: Eggenberger and Pólya [19]);

(2) Compound Poisson distribution, where the mean of the component Poisson distribution varies according to Γ distribution;

(3) Randomly distributed colonies, where the number of individuals in the colonies is distributed independently in a logarithmic distribution (Quenouille [63]); and

(4) A simple model of immigration-birth-death process, where a constant rate of immigration per unit of time and constant rates of birth and death per individual are assumed (Kendall [37]). It results in a distribution pattern identical with (3), above.

These processes roughly correspond with the main causes of animal aggregation mentioned above, though the actual biological situations are undoubtedly more complex and complicated than the simplified assumption of this model.

Burrage and Gyrisco ([11], [12]) found that the distribution of the third instar larvae of the European chafer, *Amphimallon majalis*, per 1 square foot in a 25 x 25 foot area of permanent pasture was fitted by the negative binomial distribution, but it was approximately random in each subarea of 5 x 5 feet. Similarly, the variance-mean relations of the number of the larvae of Nantucket pine tip moth, *Rhyacionia frustrana*, per tip of loblolly pine generally scattered around the Poisson expectation line if the samples from individual trees were treated separately. On the other hand, the distribution of larvae per tree followed a negative binomial distribution, which would be inclusion of trees with different tip moth population levels (Waters and Henson [85]). These cases may be regarded as examples of process (2) of this model.

The distribution of eggs of the common cabbage butterfly, *Pieris rapae*, on cabbage plants are also approximated by the negative binomial series (Kobayashi [38], [39]; and Harcourt [25]). Kobayashi [39] conducted an experiment to analyze the process generating the observed distribution and found that the female butterflies visited individual plants at random to lay eggs, but the eggs laid per visit conformed to a logarithmic distribution with mean about 1.5 (see Fig. 1). Thus the process observed is consistent with Quenouille's model (process (3)). A similar process is reported by Jones and Mollison [35] for the distribution of soil bacteria is a microscopic field.

No examples of processes (1) and (4) could be found among animal populations. Naylor [57] found that the distribution of adult flour beetle, Tribolium confusum, in flour-filled tubes arranged on the bottom of an experimental arena showed an aggregated distribution at low population densities. From the results of a more detailed analysis, he concluded that the females were distributed rather uniformly in tubes but the males tended to be attracted to the females, resulting in an aggregated distribution of mixed-sex population. He did not apply any mathematical

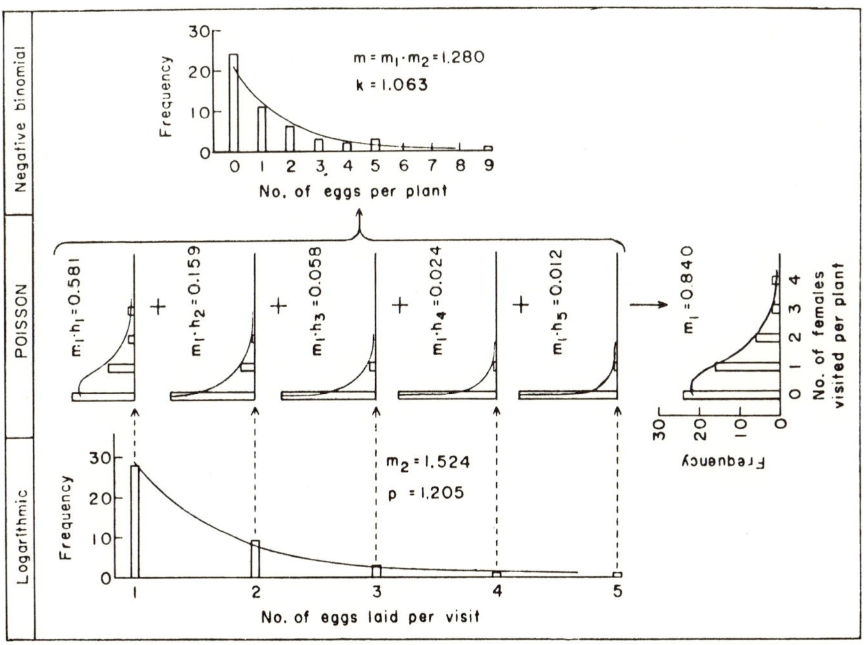

Fig. 1. Schematic diagram illustrating the distribution-generating process in the eggs of the common cabbage butterfly, Pieris rapae crucivora. The female butterflies visit cabbage plants randomly, but they lay eggs in small groups, which results in the negative binomial distribution. ([39].

model to his result, and the mechanism observed was obviously different from Pólya-Eggenberger's. However, it serves as a good example of an aggregated distribution generated from a true contagion process.

The process of population growth in some insects bears some resemblance to Kendall's model (process (4)). In such insects as the beet leafhopper, *Eutettix tenellus*, (Bowen [10]), four species of rice leafhoppers, *Nilaparvata lugens*, *Sogatella furcifera*, *Laodelphax striatellus* and *Nephotettix cincticeps* (Kuno [42], [43]), and the green peach aphid, *Myzus persicae* (Silvester and Cox [74]), the immigrant populations early in the season are known to be distributed nearly at random over the field. As the population expands through reproduction, however, their distributions become well fitted by the negative binomial series. Shiyomi and Nakamura [67] artificially settled parthenogenetic wingless females of the aphids, *Aphis maidis* and *Rhopalosiphum prunifoliae*, on barley plants according to either Poisson or uniform distributions and followed the subsequent changes in their distributions for several weeks. During the first few days after settling, the distribution per plant conformed to a Poisson distribution even when the initial distribution had been uniform, probably because plant-to-plant movement and death of aphids occurred at random. By a week later, the distribution was found to fit a negative binomial after a short transitional period during which it fit neither a Poisson nor negative binomial. The transitional period was characterized by a rapid increase of aphid numbers through reproduction and a small amount of interplant movement. After the distribution conformed to the negative binomial, the value of k increased as the population of aphids increased. Shiyomi and Nakamura attributed this change of k to the dispersal of aphids from high density plants to low density ones.

The population growth of aphids and leafhoppers is apparently different from Kendall's model in that rates of birth and death per indivdual, as well as immigration rate per unit of time, are not constant and in that the spatial expansion of colonies takes place over quadrat size. The restriction of reproductive period per generation is particularly obvious in leafhoppers. Kuno ([43]) constructed a simplified model of the distribution-generating process in rice leafhoppers, where the assumptions were a random distribution of immigrant population, a constant rate of egg-laying by females with respect to time, equal probabilities of death of adults and nymphs with respect to time and individual, and random dispersal (in the sense of Skellam [69]) in the field. The pattern of the resulting distribution and its change from generation to generation was then analyzed by means of Monte-Carlo methods. His important conclusions were that the resulting distributions in successive generations were all of the negative binomial type (Pólya-Aeppli distribution, if there was no dispersal) and that the degree of aggregation measured by C_A-index (see discussion in Section 3.2) is a function of the density of the immigrant population, dispersal ability of individuals, the rate of reproduction per generation, and the number of generations elapsed after initial invasion of the field. In the second and later generations, the degree of aggregation became higher if the reproductive rate per generation was small, which was just the opposite of the trend expected from Kendall's model.

Neyman type A distribution (Neyman [58]) was originally derived on the assumptions that egg masses are laid at random over the field, that the hatching larvae are capable of a limited movement and disperse equally in all directions, and that the number of survivors from each egg mass follows a Poisson law. Neyman type B and C distributions, as well as a generalized Neyman distribution (Beall and Rescia [4]), are based on similar but more complicated assumptions. If the number of survivors from each egg mass follows a binomial distribution instead of Poisson, the Poisson binomial distribution is derived (McGuire et al. [46]). A simpler interpretation of these distributions is that of randomly distributed colonies in which the number of individuals per colony follows either a Poisson or binomial distribution with a fixed mean (Anscombe [1]; and Skellam [70]). Similarly, if the number of individuals in colonies has either one, plus a Poisson or the negative binomial distribution, the resulting distribution is called Thomas's double Poisson (Thomas [77]) in the former or the generalized Pólya-Aeppli (Skellam [70]) in the latter.

If the observed distribution can be fitted satisfactorily to either of these models of randomly distributed colonies, we can estimate, at least theoretically, the mean size of colonies, provided that the colonies are compact and hence rarely cut through by the edge of quadrats. When the colonies are loose and often divided into portions by the quadrat edges, however, the colony size will be invariably underestimated (Pielou [60]).

Iwao (unpublished) observed the distributions of adults and egg masses of a phytophagous lady beetle, Epilachna vigintioctomaculata, in a potato field at weekly intervals. The distributions of adult females were nearly random in all of 5 quadrats of various sizes, whereas those of egg masses were apparently aggregated, particularly in small quadrats. The number of egg masses per plant on successive sampling days or those cumulated to respective dates were fitted by the negative binomial but better fitted by a Neyman type A or Thomas's double Poisson distribution. The clump size was estimated by Thomas's model, because Neyman's gave a mean, including the clumps, of size zero. The estimated mean size of clumps, around 1.20, did not increase while the mean density increased, and it was in good agreement with the average number of egg masses laid by a female per oviposition day, 1.16. Also, the marked females moved only a limited distance (average 1.6 m/wk.), and they stayed a day or longer on the same plant. From these results, the clumped nature of the egg-mass distribution appears to be produced by a mechanism resembling Thomas's double Poisson process.

According to Neyman's original assumptions, Beall [2] attempted to estimate the relative value of survival and dispersal range of some larval insects when the data were fitted by one of Neyman's distributions. The validity of these estimations, however, is questionable because of the oversimplified assumptions underlying these models, the procedure used, and the fact that these models can be derived on different assumptions.

Some highly skewed distribution of animals, such as lice on human heads, scales on citrus leaves, plankton in water samples, are known to be approximated by the logarithmic series or the discrete log-normal distributions (Cassie [13]; Spiller [72]; and Williams [86]). Cassie proposed the Poisson log-normal model on the basis of plankton data, which has a positive skewness intermediate between the negative binomial and the discrete log-normal.

It is common knowledge that an observed distribution can well be approximated by more than one mathematical model and that the model of best fit changes with the method of fitting models.

Also, whether the data are obtained by complete census of contiguous units or by sampling may influence the result. Thus, for example, Beall [2] applied Neyman's models (A, B, and C) to the distribution of the larvae of the European corn borer, Pyrausta nubilalis, per hill of corn, but Bliss and Fisher [8] and Evans [20] re-examined the same data and concluded that the negative binomial series is more appropriate than Neyman's models to describe the observed distributions. McGuire et al. [46] compared the goodness of fit of Poisson binomial, Neyman type A and the negative binomial to different sets of data of corn borer distributions per plant or per plot. They concluded that the Poisson binomial distribution was best fitted to the observed distribution when the complete census of the area was made, and that, if the data were obtained by sampling populations, the appropriate model was the negative binomial at mean density above one and the Poisson binomial at a lower density. It should be noted that in 4 of the total 9 sets of data, the observed distributions could be well approximated by all of the three distribution models tested.

Anscombe [1] pointed out that a group of two-parameter contagious distributions such as Thomas's double Poisson, Neyman type A, B and C, Pólya-Aeppli, negative binomial and discrete log-normal can be distinguished by their third and fourth factorial cumulants (see also Katti [36]). For instance, an index of skewness, $k_{[3]}m/(\sigma^2-m)^2$, where $k_{[3]}$ is the third factorial cumulant, takes the value of unity in Neyman type A, 3/2 in Pólya-Aeppli, 2 in the negative binomial, and 3 or more in discrete log-normal distributions, independent of mean and variance. Berthet and Gerard [5] and Kuno [43] applied this method to determine the distribution model appropriate for soil microarthropods (Oribatei, Acari) and for rice leafhoppers, respectively. They concluded that the observed distributions were most frequently fitted to the negative binomial series. It is noted from their figures, however, that the distribution of the index of skewness showed a considerable variation even in the same species, covering the range of all of the four mathematical distributions mentioned above (Fig. 2).

From these considerations, it is concluded that the same set of observed counts can often be described adequately by more than one mathematical model, that the distribution of the same species may vary considerably from occasion to occasion

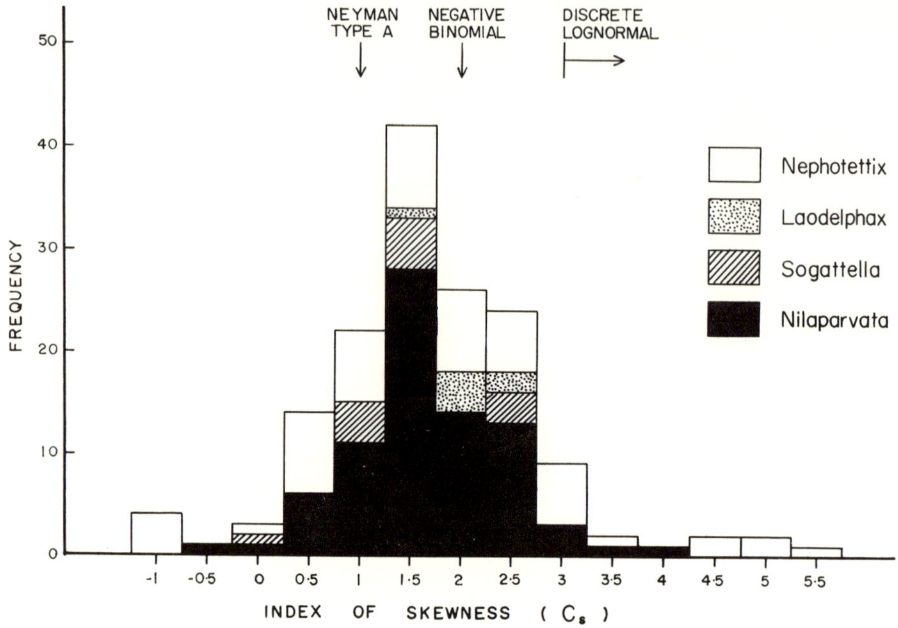

Fig. 2. Frequency distribution of the value of Anscombe's index of skewness in the distributions of 4 species of rice leafhoppers. ([43]).

to the extent that they are appropriately described by different mathematical models, and that a particular form of mathematical distribution can be derived from the different sets of underlying assumptions. Therefore, the information obtained by fitting any of the theoretical distributions hitherto proposed is rather limited, and we cannot infer the mechanism underlying the observed distribution by fitting the model unless it is combined with other biological observations.

Bliss [7] stated that in ecology fitting a mathematical distribution to the observed counts is more useful as a description of what we observe than as an analytical tool, and he recommended the negative binomial because of its widest applicability and ease of calculation of the overdispersed distributions. As Bliss stressed, the fitting of a particular distribution model to the observed data is useful for planning of sampling survey and the analysis of the data. Particularly, if sets of observed counts for a given species can reasonably be fitted by a series of negative binomial distributions with a common k, it is profitable to determine the sample size required to attain a specified degree of precision (Kuno et al. [44]; and Rojas [64]) and to develop a sequential sampling plan (e.g., Oakland [59]; Waters [82]; and Ito [30]).

However, Kobayashi [40] examined the relationship of k to the mean population density with reference to different processes leading to the negative binomial and pointed out that a common k value could be expected only in limited cases and within a limited range of mean densities. In fact, it has been reported in several animal species that the value of k tended to decrease at high mean densities (e.g., Bliss and Owen [9]; Berthet and Gerard [5]; Harcourt [25-27]; and Waters [83]). Therefore, the usefulness of calculating a common k for sampling problems appears to be rather restricted. This topic and the meaning of k as a measure of aggregation will be discussed in more detail in sections 4 and 3.2, respectively.

2.3 Truncated Distributions with Missing Zero Class

In some cases, such as the distribution of the number of larvae per gall in a gall-making insect, it is difficult to know the number of units containing no individuals. Also, generally we do not know whether all the units in a given area are equally available for the animals being studied. Bliss [7] suggested that if an observed distribution has two modes and one is at zero frequency, many units within zero class may represent contaminants from areas that differ ecologically and in which the animal would be extremely unlikely to occur.

Finney and Varley [21] found that the truncated Poisson distribution was well fitted to the frequency counts of eggs and gall-cells of the knapweed gall-fly *Urophora jaceana* per flowerhead of black knapweed. They concluded, therefore, that there was no positive evidence for the occurrence of any density-dependent mortality among larvae.

The larvae of the rice stem borer, *Chilo suppressalis*, are distributed in large clumps, each of which originates from an egg mass. At a relatively low density, individual clumps may well be identified on the distribution map. Kono [41] found that the frequency counts of the larvae per stem for each clump was well approximated by a Poisson distribution if the zero class was neglected. He considered that the number of stems containing larvae plus that of uninfested stems expected from the truncated Poisson model might represent the effective space on which a particular group of larvae orginating from an egg mass were distributed. Itô [31] studied the distribution of eggs and the first instar larvae per chestnut bud of a gall wasp, *Dryocosmus kuriphilus*. About 60 percent of the observed distributions obtained for individual trees were fitted to the truncated negative binomial and the remaining 30 percent to the truncated Poisson. The observed number of buds with no galls was much larger than the number of zeros expected from these truncated models. However, if the number of uninfested buds that were thought to be unavailable or unsuitable for female wasps at the time of their egg-laying were subtracted from the total number of uninfested buds, the observed and expected frequencies of the zero class were in general agreement with each other.

These studies propose an interesting possibility that the space available for, or that utilized by, animals might be estimated by using the truncated distribution models, though theoretical and biological background to support this possibility is yet scanty.

In the truncated Poisson distribution, the habitat units contained in the area are considered to be divided into two categories: available and unavailable. It may be reasonable to assume, however, that the suitability of individual units to animals can often be divided into many classes ranging from very suitable units to completely unsuitable (or unavailable) ones. When we assume a Poisson distribution of individuals in each class of units, the negative binomial distribution as a compound Poisson model may correspond to the case in which all the units are available but different in suitability, and the truncated negative binomial to the case in which some of the units are so unsuitable as to never be used by animals.

Cohen [15] made a simple two-parameter extension of the truncated Poisson distribution with missing zero class. The assumptions are as follows: (1) possible colony sites are distinct and countable; (2) each site has equal probability of being selected; (3) once a site is selected, one or more organisms are deposited there, and the number of individuals on a selected site is a truncated Poisson variable with missing zero class;

and (4) there is no migration between colonies. This gives another interpretation when the observed counts are fitted by the truncated Poisson, where all of the units are assumed to be equally available for the animals.

2.4 Underdispersed (Uniform) Distributions

Underdispersed or uniform distributions are rarely found in animal populations under natural conditions, and the mathematical models for them have not yet been studied in depth. A simple underdispersed distribution is the positive binomial series. It is expected when any sampling unit has an equal probability of being occupied by an individual but the maximum number of individuals a unit could contain is limited physically or in a way that the presence of one individual in a unit decreases the chances of other individuals falling in the same unit for some reason (Student [73]; Torii [78]; and Cassie [13]). As far as I know, the fitting of this model to animal distribution has never been made. However, perhaps it can be applicable for some cases where a simple form of competition between individuals limits the maximum number possible in discrete habitat units. The expected distribution is given by the expansion of $(q+p)^{k'}$, where k' is the 'capacity' or the maximum number possible per unit, p the probability of any given point being occupied by an individual, and $q = 1-p$. Since $m = k'p$ and $\sigma^2 = k'pq$, the replacement of m and σ^2 by sample values $\bar{x}$ and s^2 leads to a simple estimate of the capacity: $k' = \bar{x}^2/(\bar{x} - s^2)$.

In the azuki bean weevil, Callosobruchus chinensis, the distribution of adults emerged per bean showed an underdispersed distribution, where the maximum number per bean was limited to about 11 because of the competition among larvae (Utida [79]). The value of k' estimated by the above method is always smaller than 11, indicating that the observed distribution is more uniform than the distribution expected from the positive binomial series.

The underdispersed distribution will be discussed further in relation to the regression method for the analysis of distribution pattern.

3. MEASURES OF AGGREGATION FOR INDIVIDUAL DISTRIBUTION

Many indices of dispersion have been proposed in order to analyze animal and plant distributions without any *a priori* assumption of the type of distribution (see Greig-Smith [24]). Blackman [6], Utida et al. [80] and many others used the variance-mean ratio ($s^2/\bar{x}$, and David and Moore [17] used $s^2/\bar{x} - 1$; but these and many other indices have a serious defect in that the value of the index is affected by the level of mean density (Morisita [51]). In this section I shall mention only a few indices that are considered to be most useful.

3.1 I_δ-index

I_δ-index proposed by Morisita [51] is expressed by

$$I_\delta = \frac{q \sum_{i=1}^{q} x_i(x_i-1)}{T(T-1)} ,$$

where q is the number of sampling units, x_i the number of individuals in the ith unit, and T the total number of individuals in the sample ($= \sum_{i=1}^{q} x_i$). It is based on the unbiased estimate of Simpson's [68] measure of diversity, $\hat{\lambda} = \Sigma x_i(x_i-1)/T(T-1)$, which is the probability that two individuals chosen at random and independently from the population will be in any one of the groups into which the individuals of an infinite population are divided.

I_δ takes the value unity when the distribution is random, larger than unity for aggregated distributions, and smaller than unity for underdispersed distributions. Approximate relationships of I_δ to variance-mean ratio and the negative binomial's k are shown as $s^2/\bar{x} = (I_\delta-1)\bar{x}+1$ and $1/k \doteqdot E(I_\delta)-1$, respectively (Morisita [52]).

By using the change of I_δ value with the increase of quadrat size, it is possible to detect the more detailed nature of the observed distribution such as the approximate size of clumps and the pattern of intra-clump distribution. (Fig. 3) Although I_δ was first derived as a statistic, Morisita ([53], [54]) made the distinction between I_δ of the population and its estimate, and he derived several formulae basic for the application of I_δ to sampling problems.

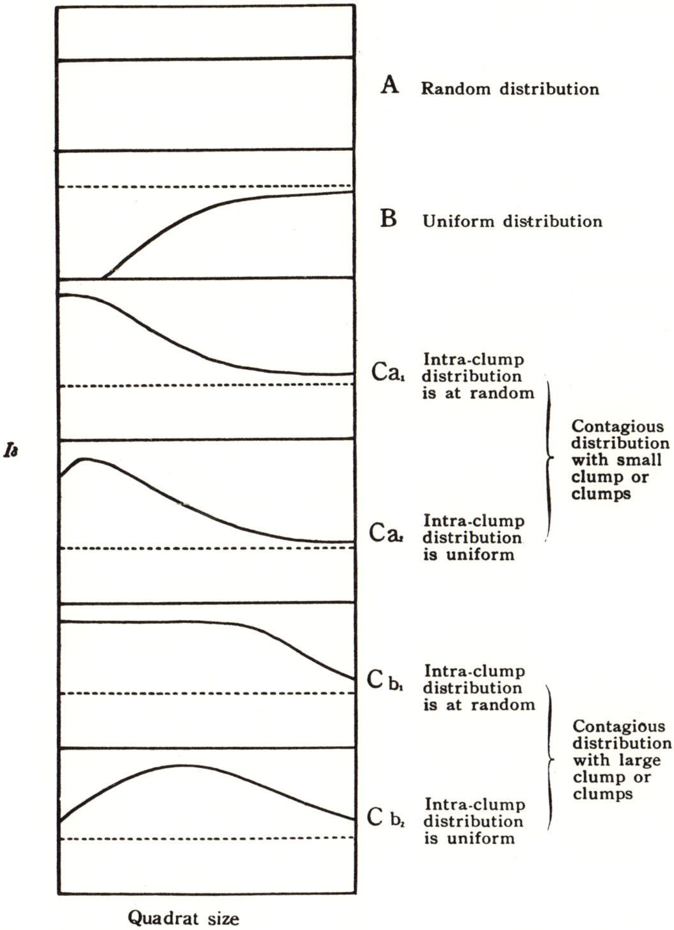

Fig. 3. Schematic representation of the relation of I_δ to quadrat size in various types of spatial distributions. The broken line in each figure indicates the Poisson expectation line (I_δ = 1). ([51]).

3.2 C_A-index ($\approx 1/k$)

Waters [83] stressed that k, the exponent of the negative binomial, would be a good measure of aggregation because of this wide applicability and flexibility of this distribution. Since k becomes infinite as the distribution becomes random, it is preferable to use $1/k$, which takes the value zero for a random distribution and increases as overdispersion increases (Cassie [13]).

Recently, Kuno [43] clearly demonstrated that $C_A = (\sigma^2 - m)/m^2$, which corresponds to $1/k$ in the negative binomial, can be an adequate measure of the degree of aggregation in the population, regardless of the type of the distribution: since the variance shows the variation around the mean of the number of individuals in quadrats, σ^2/m^2 indicates the relative magnitude of variation independent of the mean. It takes the value $(1-1/Q)/m$ for a random distribution, where Q is the total number of quadrats in the population. Subtracting $(1-1/Q)/m$ from σ^2/m^2, we have

$$C'_A = \frac{\sigma^2}{m^2} - \frac{1}{m}\left(1-\frac{1}{Q}\right),$$

as a relative measure of aggregation, which takes the value zero for a random distribution. If $Q >> 0$ as is usually the case, it is reduced to

$$C_A = \frac{\sigma^2 - m}{m^2}\ .$$

If the individuals are removed at random with a given probability p from the population having the distribution f(x), the subsequent distribution g(x) can be shown by

$$g(x) = \sum_{m=0}^{\infty} f(n)\binom{n}{x} q^x p^{n-x},$$

where $q = 1-p$. The mean and variance of the initial population, m and σ^2, change to mq and $\sigma^2 q^2 + pqm$, respectively, regardless of the form of f(x). Namely, the value of C_A does not change after random removal of individuals, demonstrating that C_A is not affected by the level of mean density.

Kuno suggested that an approximately unbiased estimate of C_A can be given by

$$\hat{C}_A = \frac{s^2 - \bar{x}}{\bar{x}^2 - s^2/n} \qquad \text{(n: number of sample units)}$$

which is the equation given by Bliss and Owen [9] for estimating $1/k$ in the negative binomial distribution.

3.3 Mean Crowding and Patchiness

Lloyd [45] proposed an interesting parameter, "mean crowding," which is defined as the mean number of other individuals per individual per quadrat and expressed by

$$\overset{*}{m} = \frac{\sum_{i=1}^{Q} x_i(x_i - 1)}{\sum_{i=1}^{Q} x_i} = m + (\frac{\sigma^2}{m} - 1),$$

where Q is the total number of quadrats in the population. His intention was to measure the average degree of crowing experienced by individuals in a population (i.e., disoperative influences from other individuals), so that this parameter in its original sense can be applied properly only to freely moving, nongregarious animals that are living in continuous and uniform habitats. Iwao [32], however, pointed out that the concept of mean crowding as defined above could generally be applied for any kind of organisms and for any size of quadrat unit, without special implication on the effect of crowding.

The sample estimate of mean crowding, $\overset{*}{x}$, can be obtained by replacement of m and σ^2 by their sample estimates $\bar{x}$ and s^2 in the above equation:

$$\overset{*}{x} = \bar{x} + (\frac{s^2}{\bar{x}} - 1),$$

though it may be biased to a certain extent. Lloyd derived serveral equations for estimating $\overset{*}{m}$ when the distribution follows the negative binomial series. Among them, the estimation by moments is given by

$$\overset{*}{x} = \bar{x} + (\frac{s^2}{\bar{x}} - 1)\left(1 + \frac{s^2}{q\bar{x}^2}\right),$$

which would be applicable for most of the overdispersed distributions.

For the ratio of mean crowding the mean density, we get the relation

$$\frac{\overset{*}{m}}{m} = 1 + \frac{\sigma^2 - m}{m^2} = 1 + C_A \approx I_\delta .$$

Lloyd called this ratio patchiness. Thus the three indices, I_δ, C_A, and $\overset{*}{m}/m$, are considered to be essentially identical, though they are derived on the basis of different theoretical considerations.

The analysis of animal distribution by using one of these indices can yield biologically interesting results, if it is used in conjunction with other methods. For example, Nakasuji et al. [56], using I_δ, studied the distributions of adults and egg masses of three species of plant bugs, Nezara viridula, Scotinophara lurida and Acanthocoris sordidus. The distribution patterns of these species are summarized in Table 1. To explain the observed differences, Nakasuji et al. also made analyses of the spatial association between both sexes of adults, that between females and egg masses, and movement of marked adults.

Table 1. The distribution of three species of plant bugs. (From [56].)

Species	Habitat	Distribution pattern of		
		Males	Females	Egg Masses
N. viridula	Paddy field	contagious	random	random
S. lurida	Paddy field	contagious	random	rather contagious
A. sordidus	Potato field	random	contagious	contagious

In N. viridula, female adults moved around actively and laid their egg-masses successively at rather long intervals of time, so that both females and egg masses were distributed nearly at random over the field. The males were less active and tended to be attracted to females, resulting in a contagious pattern of their distribution. In S. lurida, females were distributed randomly, but they laid egg masses at short intervals, so that the distribution of egg masses became somewhat contagious and it overlapped spatially with the female's distribution. In A. sordidus, females were sluggish and remained near the edge of the field after invading there and laid egg masses successively at short intervals, which led to contagious distributions of both females and egg masses. From the change of I_δ value with the increase of unit size, the clump sizes of female and egg-mass distribution in A. sordidus were suggested to be in the range of 4 plants of potatoes, which was consistent with the mean dispersal range of marked females.

4. REGRESSION ANALYSIS OF AGGREGATION

In the spatial distribution problem, we are concerned with two different aspects of aggregation pattern: one, whether or not the individuals are distributed in groups or colonies; and the other, whether or not the distribution pattern of such basic units (individuals or groups of individuals) is contagious. These two aspects cannot usually be distinguished by means of a single index of aggregation or of fitting a mathematical distribution to the observed counts.

Since animals, as well as plants, frequently live in tight or loose aggregations because of their gregarious habits or their limited dispersal ability, coupled with mass deposition of eggs or young, it is important to distinguish the two aspects of aggregation and also to separate the apparent and true changes of animal distribution pattern with change of their living conditions.

Iwao [32] proposed a regression method for the analysis of distribution patterns. The method is based on the the regression of mean crowding on mean density over a range of different levels of densities. The regression proves to be rectilinear in various mathematical distributions. Namely, the relationship of mean crowding ($\overset{*}{m}$) to mean density (m) can be shown by

$$\overset{*}{m} = \alpha + \beta m,$$

where α is the intercept on the $\overset{*}{m}$ axis and β the regression coefficient. The values of α and β are characteristic of respective distributions as shown in Table 2.

In the models of randomly distributed colonies, it can easily be understood that α equals the mean crowding for the frequency distribution of the number of individuals per colony. Generally speaking, α corresponds to the mean number per individual of other individuals in the same unit when the mean density becomes infinitesimal. Therefore, it will take the value zero for distributions where single individuals are the basic units and it will take a positive value in any kind of colonial distribution. The value of β equals unity when the individuals or the colonies (with a fixed mean) are distributed at random, larger than unity when they are distributed contagiously, and smaller than unity (minimum zero) when they are underdispersed.

In the completely uniform distribution, α and β take values

Table 2. The relation of α and β to the type of spatial distribution.

Distribution pattern	The value of α	The value of β	Possible Explanation of underlaying mechanism*	Corresponding mathematical models
Random	0	1	Random	Poisson
Over-dispersion	>0	1	Randomly distributed colonies (mean colony size: fixed) True contagion	Poisson binomial, Neyman A, Thomas's, Negative binomial (Quenouille's), Pólya-Aeppli Negative binomial (Pólya-Eggenberger's type with a common d = m/k) $\alpha = \sigma^2/m-1$
	0	>1	Distributions with a definite tendency of aggregation (in sense of constant I_δ or $\overset{*}{m}/m$)	Negative binomial series with a common k ($\beta = 1+1/k$)
	>0	>1	Contagiously distributed colonies (mean colony size: fixed); Many other processes may be included in this category.	No mathematical models (e.g., distributions of colonies follow the negative binomial series with a common k)
Under-dispersion	0	<1	Maximum number of individuals possible in a unit is limited in some way	Positive binomial series with a common k' ($\beta = 1-1/k'$)
	0 -1	0 for $m \leq 1$ $\dotplus 1$ for $m \geq 1$	Completely uniform distribution	

* Does not include all the possible processes.

from zero up to the mean density 1.0. For the range of higher densities, however, β approximately equals unity (some waving of the regression is produced by the discrete nature of the distribution), and the value of α extrapolated from this range is -1.0. This kind of distribution pattern would be expected when a strong repulsion occurs, such as territorial behavior or the avoidance of the place conditioned by other individuals.

Thus, α indicates the tendency of aggregation (positive contagion) or repulsion (negative contagion) mostly inherent to the species, which forms a basic component of distribution. It is therefore termed "index of basic contagion." On the other hand, β defines the spatial pattern of habitat utilization with changes in mean density, which may be related with heterogeneity of the environment in many cases. It is called "density-contagiousness coefficient."

Since α and β are concerned with different aspects of distribution, any combination of these two parameters would be possible within the range of $-1 \leq \alpha$ and $0 \leq \beta$.

The estimates of α and β can be obtained using $\overset{*}{x}$ and $\bar{x}$ in a series of observed counts. The linearity of the regression of $\overset{*}{x}$ on $\bar{x}$ has been demonstrated in most of the animal distributions tested (Iwao [32]). Several interesting points revealed are summarized as follows:

(1) We can distinguish three types of contagious distributions, in most of which composite distributions in a series can be approximated by the negative binomial distributions: the distributions with $\hat{\alpha} \doteqdot 0$ and $\hat{\beta} > 1$, which corresponds to the negative binomial series with a common k (e.g., spruce budworm larvae, Bliss [7]); the distributions with $\hat{\alpha} > 0$ and $\hat{\beta} \doteqdot 1$, which corresponds to models of randomly distributed colonies (e.g., eggs of *Pieris rapae*, Kobayashi [39]; egg masses of *Epilachna vigintioctomaculata*, also discussed in section 2.2); and the distributions with $\hat{\alpha} > 0$ and $\hat{\beta} > 1$, which include many examples where a decreasing trend of 1/k with increasing density is observed (e.g., wireworm larvae, Bliss and Owen [9]; and several species of soil Oribatids, Berthet and Gerard [5]);

(2) A single linear regression can often be fitted to a series of observed counts, where the mathematical model of best fit changes as mean density increases or no model can be fitted to some of them (e.g., corn borer larvae, McGuire *et al.* [46]);

(3) A single linear regression can often be fitted to the observed counts that are taken from different sources with regard to time, place, or stratum (e.g., pea aphid, Forsythe and Gyrisco [22]; bamboo gall-wasp, Aiolomorphus rhopaloides, Iwao [unpublished]);

(4) A single linear regression may still be fitted even when the distribution pattern changes from underdispersion to over-dispersion with a change in mean density (e.g., $\hat{\alpha} < 0$ & $\hat{\beta} > 1$: eggs of azuki bean weevil, Iwao [32]; $\hat{\alpha} > 0$ & $\hat{\beta} < 1$: adults of flour beetle, Naylor [57]).

Thus the method can be applied for a wide variety of actual distributions of animals. If a single regression can be fitted to a set of distributions of a given species, there seems to be no reason to assume any change in the mechanism of distribution with changes in mean density (discussed further two pages hence). From the relation $\overset{*}{m}/m = \alpha/m+\beta$, it is obvious that the value of the single index of aggregation decreases with increasing density if $\alpha \neq 0$, without any change of distribution mechanism.

If the values of α and β are estimated for a given species and prove to be stable at least under certain conditions, the sample size needed to attain a specified precision in simple random sampling can easily be determined for a given level of mean density. It is given by

$$q = \frac{t^2}{D^2}\left(\frac{\hat{\alpha}+1}{\bar{x}} + \hat{\beta} - 1\right),$$

where t is the Student's statistic and $D = ts_{\bar{x}}/\bar{x}$ ($s_{\bar{x}}$: standard deviation of the mean) (Iwao and Kuno [33]). Application of the method to stratified sampling and two-stage sampling is also possible (Iwao and Kuno [34]).

The method of transformation of data to stabilize variance can be derived also. The general equation for the transformation of original counts x_i to $f(x_i)$ is given by

$$f(x_i) = c\int \frac{1}{\sqrt{(\alpha+1)x_i+(\beta-1)x_i^2}}\,dx.$$

From this, a set of equations appropriate for various combinations of α and β can be deduced. Among others, for $\alpha > -1$ and $\beta > 1$, we have

$$f(x_i) = \sinh^{-1}\sqrt{\frac{\beta-1}{\alpha+1}x_i} = \log_e\left(\sqrt{\frac{\beta-1}{\alpha+1}x_i} - \sqrt{\frac{\beta-1}{\alpha+1}x_i+1}\right),$$

which would have widest applicability, because most of the biological distributions fall into this category. This leads to Beall's transformation $f(x_i) = \sinh^{-1} \sqrt{x_i/k}$, if the distributions follow the negative binomial series with a common k (i.e., $\alpha = 0$, $\beta = 1 + 1/k$). Also, when $\alpha > -1$ and $\beta = 1$, the general equation becomes $f(x_i) = \sqrt{x_i}$, which includes the Poisson series ($\alpha = 0$, $\beta = 1$) (Iwao and Kuno [33]).

The method also will be useful as a tool for analyzing the behavior of individuals, the process of specio-temporal utilization of habitat by a population, and the distribution of density-related and unrelated mortality processes, etc. These topics will be discussed in a separate paper (Iwao and Kuno, [34]). Some of the important conclusions are summarized as follows:

(1) If a proportion p of the individuals in a population are randomly removed, the mean crowding and the mean density of the initial population decreases at the same rate p. Then, when a series of populations that have common values of α_1 and β_1 are similarly affected by a constant probability p, the resulting series of the populations would have $\alpha_2 = (1-p)\alpha_1$ and $\beta_2 = \beta_1$ (if $\alpha_1 = 0$, the regression does not change at all). This corresponds to the operation of a density-independent mortality process. In a series of colonial distributions that are fitted to a single regression, the random removal of colonies at a constant rate p does not change the values of α and β (all-or-none type random mortality upon colonies).

(2) If the population increases by a given ratio c (>1), the change occurs in the direction just opposite to case (1). Namely, the mean crowding of the original population will increase by c times. Then, in a series of populations, α becomes c times larger, but β remains the same. Also, adding new colonies by the same rule does not change both α and β. If the individuals in every unit increase deterministically at a constant rate c', $\overset{*}{m}$ and $(\overset{*}{m}+1)$ will increase by c' times. Then, if a series of populations equally change at a rate c', $(\alpha+1)$ becomes $c'(\alpha+1)$ but β does not change.

Thus, the method has a wide applicability. It is an advantage of this method that it is mathematically related in a strict sense with various theoretical distributions as well as mean and variance. If only one or a few sets of distribution data are available, we cannot estimate α and β; we can only get estimates of means and mean crowdings. Even in such a case, the method will give us more information than other single indices, because mean crowding itself is a meaningful parameter of the population.

Taylor ([75], [76]) believed that the relation between mean ($\bar{x}$) and variance (s^2) in the frequency counts of many species of animals and plants would follow a power law: $s^2 = a\bar{x}^b$. Then a linear regression can be fitted to the relation of $\log(s^2)$ to $\log \bar{x}$ in a series of distributions for a given species. He pointed out that a is largely a sampling and computing factor but b could be a true index of aggregation characteristic of the species. In Iwao's regression method, the variance σ^2 is expressed by $(\alpha+1)m+(\beta-1)m^2$. This suggests that the strict linearity of the regression of $\log(s^2)$ on $\log \bar{x}$ is only expected when $\alpha = -1$ or $\beta = 1$. In fact, such theoretical distributions as the negative binomial series with a common k and such biological distributions for which β is considerably larger than unity show apparent curvilinear relations of $\log(s^2)$ on $\log \bar{x}$. Also, even when Taylor's law holds good, as in the cases of Poisson series and the models of randomly distributed colonies, the value of b does not provide any information on the important difference between these types of distributions. The system of transformation of data given by Taylor [75] and Healy and Taylor [28] also is less adequate than the method given by Iwao and Kuno [33]. It seems, therefore, that Taylor's method lacks a sufficient theoretical background, and it provides only an approximate, empirical "law" in the spatial distribution of organisms.

5. CONCLUDING REMARKS

By way of conclusion, the present status and the perspective of spatial distribution problems in animal population ecology will be briefly mentioned in turn.

5.1 Description of Spatial Distribution Pattern of a Species

As mentioned previously, the description of distribution pattern of animals by using one of the mathematical models available is only approximate and superficial, and the analysis by an aggregation index may often be sufficient for the information required. In either case, we need to have some additional information if the mechanism leading to the observed spatial pattern is to be analyzed.

The method based on the relation of mean crowding to mean density, first proposed by Lloyd [45] and extended to a regression method by Iwao [32], provides a comprehensive way to describe the distribution pattern of a species. The two parameters, the index of basic contagion (α) and the density-contagiousness coefficient (β) have rather definite biological meanings. Basic patterns generated from various mathematical models are included as special cases in the system. It seems, therefore, that this regression method is the best for describing the general characteristics of the distribution pattern of a species in the habitat units of a given size. In connection with this method, we can use an appropriate mathematical distribution if we need a more detailed description of the form of distribution.

5.2 Sampling Techniques and Transformation of Data

It can be said that the basic problems involved in sampling and the transformation of data are now solved more generally and systematically by Iwao's regression method than by application of mathematical distributions. Although the sequential sampling method is yet applicable only when a series of observed counts follows one of a few distribution models, it seems possible to extend its applicability by developing the equations on the basis of this regression method (Iwao [32]). Using α and β, Kuno [43a] has developed a method of sequential sampling that involves a flexible sample size with a sufficient number of sample units taken to obtain the estimated mean at a predetermined degree of its precision, which seems to be more useful than the ordinary sequential method for ecological studies.

5.3 Analysis of Ecological Processes

Some useful information on the ecological processes going on in a population would be obtained by analyzing the change of spatial distribution during successive intervals of time. For this purpose, first of all we should find out definite relationships between the spatial pattern and the ecological variables. The change of distribution pattern in relation to the population density is of special importance, because the regulation of population size by density-dependent processes is a central problem in animal population ecology. In an early attempt along this line, Utida *et al*. [80] used the variance-mean ration to detect the density-dependent change (through mortality and/or dispersal) in the distribution pattern of the common cabbage butterfly, Pieris rapae, during development. Morisita [52] criticized the use of variance-mean ratio because it is affected by the level of mean density, and he recommended I_δ as a better criterion. However, even I_δ or allied indices may change values in the process of spatial expansion of initial aggregates in the course of development or with changes in the population density of animals of the same stage, without operation of any density-dependent process. Therefore, it is not usable as a general measure for this purpose. Lloyd's [45] similar attempt using changes in mean (m), mean crowding ($\overset{*}{m}$) and the ratio $\overset{*}{m}/m$ or $(\overset{*}{m}+1)/m$ also, at least in part, suffers from the same criticism (Iwao [1968]; see also section 4). The regression method based on the m-$\overset{*}{m}$ relation appears to be most promising as a means of detecting density dependence, as mentioned briefly in the foregoing section.

5.4 Use of Mathematical Distribution for the Construction of Population Models

An obvious merit of a mathematical distribution is that it defines the expected frequency distribution precisely for given values of its parameters. Then, we can construct a realistic population model by taking a spatial distribution model widely applicable to biological populations as a basic component. For example, the rate of attack by a predator on prey may surely be influenced not only by prey density, but also by its distribution pattern; and hence the model of predator's attack based on the negative binomial distribution of prey population would be more realistic and versatile than that based on the Poisson distribution of it. This kind of simulation model on predator-prey relations is being developed by Holling and his colleagues (Holling [29]).

5.5 Towards More Informative Distribution Models

It is very profitable if we can estimate some ecologically important parameters, such as birth and death rates, index of dispersal, etc., through the fitting of a mathematical model to the observed distribution. To meet such a requirement, the mathematical model should be constructed on the basis of realistic biological assumptions. A series of Neyman's contagious distributions may be regarded as a classical example of this kind of attempt, though it is based on oversimplified assumptions and proves to be unsuccessful for this purpose.

Since spatial distribution of organisms at a given time is the result of complex biological processes that were going on in the population at that moment, the analyses and mathematical formulation of the important processes determining the distribution would be needed to construct more informative distribution models. On the process of dispersal, Skellam [69] and Watanabe et al. [81] developed essentially similar models of random dispersal, which resulted in the bivariate normal distribution. They are somewhat related to Morisita's [49] empirical equation of animal dispersal from a point. Shiyomi ([65], [66]) presented a model of plant-to-plant movement of apterous insects by the convolution of Poisson with binomial distributions. Murai [55] considered mathematically the change of distribution pattern of animals between two successive stages due to competition, by assuming the simplified models of scramble and contest type competition (Fujii [23]). Obviously, more analytical studies on these and other aspects are necessary.

As an important step to combine the respective components into an integrated distribution model, the studies on the distribution by the method of simulation, as has been done by Kuno ([43]; see section 2.2) will be very valuable.

ACKNOWLEDGEMENT

I am deeply indebted to Dr. Eizi Kuno of our Laboratory for his critical reading of the manuscript.

REFERENCES

[1] Anscombe, F. J. 1950. Sampling theory of the negative binomial and logarithmic series distributions. Biometrika 37:358-82.

[2] Beall, G. 1940. The fit and significance of contagious distributions when applied to observations on larval insects. Ecology 21:460-74.

[3] ———. 1942. The transformation of data from entomological field experiments so that the analysis of variance becomes available. Biometrika 32:243-62.

[4] ———, and Rescia, R. R. 1953. A generalization of Neyman's contagious distributions. Biometrics 9:354-86.

[5] Bethet, P., and Gerard, G. 1965. A statistical study of microdistribution of Oribatei (Acari). Part I. The distribution pattern. Oikos 16:214-27.

[6] Blackman, G. E. 1942. Statistical and ecological studies on the distribution of species in plant communities. Ann. Bot. London, N.S. 6:351-66.

[7] Bliss, C. I. 1958. The analysis of insect counts as negative binomial distributions. Proc. 10th Intern. Congr. Ent. 1956, 2:1015-32.

[8] ———, and Fisher, R. A. 1953. Fitting the negative binomial distribution to biological data. Biometrics 9:176-200.

[9] ———, and Owen, A. R. G. 1958. Negative binomial distributions with a common k. Biometrika 45:39-58.

[10] Bowen, M. F. 1947. Population distribution of the beet leafhopper in relation to experimental field-plot layout. J. Agric. Res. 75:259-78.

[11] Burrage, R. H., and Gyrisco, G. G. 1954. Distribution of third instar larvae of the European chafer and the efficiency of various sampling units for estimating their populations. J. Econ. Ent. 47:1009-14.

[12] ———, and ———. 1956. The transformation of counts of Eurpoean chafer larvae for analysis of variance. J. Econ. Ent. 49:179-82.

[13] Cassie, M. 1962. Frequency distribution models in the ecology of plankton and other organisms. J. Anim. Ecol. 31:65-92.

[14] Clark, P. J., and Evans, F. C. 1954. Distance to nearest neighbor as a measure of spatial relationships in populations. Ecology 35:445-53.

[15] Cohen, A. C. Jr. 1960. An extension of a truncated Poisson distribution. Biometrics 16:446-50.

[16] Cole, L. C. 1946. A study of the cryptozoa of an Illinois woodland. Ecol. Monogr. 16:49-86.

[17] David, F. N., and Moore, P. G. 1954. Notes on contagious distributions in plant populations. Ann. Bot. London, N.S. 18:47-53.

[18] Eberhardt, L. L. 1967. Some developments in 'distance sampling.' Biometrics 23:207-16.

[19] Eggenberger, F., and Pólya, G. 1923. Uber die statistick verketteter Vörgange. Z. Angew. Math. u. Mechanik 3:279-89.

[20] Evans, D. A. 1953. Experimental evidence concerning contagious distributions in ecology. Biometrika 40:186-211.

[21] Finney, D. J., and Varley, G. C. 1955. An example of the truncated Poisson distribution. Biometrics 11:387-94.

[22] Forsythe, H. Y., Jr., and Gyrisco, G. G. 1963. The spatial pattern of the pea aphid in alfalfa fields. J. Econ. Ent. 56:104-7.

[23] Fujii, K. 1965. A statistical model of the competition curve. Res. Popul. Ecol. 7:118-25.

[24] Greig-Smith, P. 1964. Quantitative Plant Ecology. 2nd ed. Butterworths, London.

[25] Harcourt, D. G. 1961. Spatial pattern of the imported cabbageworm, Pieris rapae (L.) (Lepidoptera:Pieridae), on cultivated Cruciferae. Can. Ent. 93:945-52.

[26] ———. 1963. Population dynamics of Leptinotarsa decemlineata (Say) in eastern Ontario. I. Spatial pattern and transformation of field counts. Can. Ent. 95:813-20.

[27] ———. 1965. Spatial pattern of the cabbage looper, Trichoplusia ni, on Crucifers. Ann. Ent. Soc. America 58:89-94.

[28] Healy, M. J. R., and Taylor, L. R. 1962. Tables for power-law transformations. Biometrika 49:557-9.

[29] Holling, C. S. 1968. The tactics of a predator. In 'Insect Abundance' (ed. by T. R. E. Southwood). Symp. Roy. Ent. Soc. London No. 4, pp. 47-58.

[30] Itô, Y. 1962. Distribution of the overwintering arrowhead scale, Prontaspis yanonensis Kuwana, on the Satsuma Orange leaves. Jap. J. Appl. Ent. Zool. 6:183-9. (In Japanese with English summary)

[31] ———. 1967. Population dynamics of the chestnut gall-wasp, Dryocosmus kuriphilus Yasumatsu (Hymenoptera: Cynipidae). IV. Further analysis of the distribution of eggs and young larvae in buds using the truncated negative binomial series. Res. Popul. Ecol. 9:177-91.

[32] Iwao, S. 1968. A new regression method for analyzing the aggregation pattern of animal populations. Res. Popul. Ecol. 10:1-20.

[33] ————, and Kuno, E. 1968. Use of the regression of mean crowding on mean density for estimating sample size and the transformation of data for the analysis of variance. Res. Popul. Ecol. 10:210-4.

[34] ————, and ————. 1970. An approach to the anslysis of aggregation pattern in biological populations. In Statistical Ecology, Vol. I., Spatial Patterns and Statistical Distributions, Eds. G. P. Patil, E. C. Pielou, and W. E. Waters, Penn. State U. Press (to appear).

[35] Jones, P. C. T., and Mollison, J. E. 1948. A technique for the quantitative estimation of soil microorganisms. J. Gen. Microbiol. 2:54-69.

[36] Katti, S. K. 1966. Interrelations among generalized distributions and their components. Biometrics 22:44-52.

[37] Kendall, D. G. 1948. On some modes of population growth leading to R. A. Fisher's logarithmic series distribution. Biometrika 35:6-15.

[38] Kobayashi, S. 1965. Influence of parental density on the distribution pattern of eggs in the common cabbage butterfly, Pieris rapae crucivora. Res. Popul. Ecol. 7:109-17.

[39] ————. 1966. Process generating the distribution pattern of eggs of the common cabbage butterfly, Pieris rapae crucivora. Res. Popul. Ecol. 8:51-61.

[40] ————. 1968. Population density and degree of aggregation, with special reference to a common k in the negative binomial distributions. Jap. J. Ecol. 18:120-4. (In Japanese with English summary).

[41] Kono, T. 1953. Basic unit of population observed in the distribution of the rice-stem borer, Chilo simplex, in a paddy field. Res. Popul. Ecol. 2:95-105. (In Japanese with English summary).

[42] Kuno, E. 1963. A comparative analysis of the distribution of nymphal populations of some leaf- and planthoppers on rice plant. Res. Popul. Ecol. 5:31-43.

[43] ————. 1968. Studies on the population dynamics of rice leafhoppers in a paddy field. Bull. Kyushu Agric. Expt. Sta. 14:131-246. (In Japanese with English summary.)

[43a] ————. 1969. A new method of sequential sampling to obtain the population estimates with a fixed level of precision. Res. Popul. Ecol. 11:127-36.

[44] ————, Yamamoto, S.; Satomi, H.; Ōuti, Y.; and Okada, T. 1963. On the assessment of insect populations in a large area of paddy field based on the negative binomial distribution. Proc. Assoc. Plant Prot. Kyushu 9:33-6. (In Japanese with English summary).

[45] Lloyd, Monte. 1967. 'Mean crowding.' J. Anim. Ecol. 36:1-30.

[46] McGuire, J. U.; Brindley, T. A.; and Bancroft, T. A. 1957. The distribution of European corn borer larvae *Pyrausta nubilalis* (Hbn.) in field corn. Biometrics 13:65-78.

[47] Morisita, M. 1950. Dispersal and population density of a waterstrider, *Gerris lacustris* L. Contr. Physiol. & Ecol., Coll. Sci., Kyoto Univ. No. 65, 1-149. (In Japanese).

[48] ————. 1954. Estimation of population density by spacing method. Mem. Fac. Sci. Kyushu Univ., Ser. E (Biol.) 1:187-97.

[49] ————. 1954. Dispersion and population pressure. Experimental studies on the population density of an ant-lion, *Glenuroides japonicus* M'L. (II). Jap. J. Ecol. 4:71-9. (In Japanese with English summary).

[50] ————. 1957. A new method for the estimation of density by the spacing method applicable to non-randomly distributed populations. Physiol. & Ecol. (Kyoto) 7:134-44. (In Japanese with English summary).

[51] ————. 1959. Measuring of the dispersion of individuals and analysis of the distributional patterns. Mem. Fac. Sci. Kyushu Univ., Ser. E (Biol.) 2:215-35.

[52] ————. 1962. I_δ-index, a measure of dispersion of individuals. Res. Popul. Ecol. 4:1-7.

[53] ————. 1964. Application of I_δ-index to sampling techniques. Res. Popul. Ecol. 6:43-53.

[54] ————. 1965. A revision of the methods for estimating population values of the index of dispersion in the I_δ-method. Res. Popul. Ecol. 7:126-8.

[55] Murai, M. 1966. A statistical consideration on the change of the distribution pattern between two successive stages. Res. Popul. Ecol. 8:8-13.

[56] Nakasuji, F.; Hokyo, N.; and Kiritani, K. 1965. Spatial distribution of three plant bugs in relation to their behavior. Res. Popul. Ecol. 7:99-108.

[57] Naylor, A. F. 1959. An experimental analysis of dispersal in the flour beetle, Tribolium confusum. Ecology 40:453-65.

[58] Neyman, J. 1939. On a new class of 'contagious' distributions, applicable in entomology and bacteriology. Ann. Math. Stat. 10:35-57.

[59] Oakland, G. B. 1950. An application of sequential analysis to white fish sampling. Biometrics 6:59-67.

[60] Pielou, E. C. 1957. The effect of quadrat size on the estimation of the parameters of Neyman's and Thomas's distributions. J. Ecol. 45:31-47.

[61] ———. 1959. The use of point-to-plant distances in the study of pattern of plant populations. J. Ecol. 47:607-13.

[62] ———. 1960. A single mechanism to account for regular, random and aggregated population. J. Ecol. 48:575-84.

[63] Quenouille, M. H. 1949. A relation between the logarithmic, Poisson and negative binomial series. Biometrics 5:162-4.

[64] Rojas, B. A. 1964. La binomial negativa y la estimación de intensidad de plagas en el suelo. Fitotecnia Latinamer. 1:27-36 (cited from [71], below).

[65] Shiyomi, M. 1967. A model of the plant-to-plant movement of aphids. I. Description of the model. Res. Popul. Ecol. 9:53-61.

[66] ———. 1968. A model of the plant-to-plant movement of aphids. II. Estimation of parameters and the application. Res. Popul. Ecol. 10:105-14.

[67] ———, and Nakamura, K. 1964. Experimental studies on the distribution of the aphid counts. Res. Popul. Ecol. 6:79-87.

[68] Simpson, E. H. 1949. Measurement of diversity. Nature (Lond.) 163:688.

[69] Skellam, J. G. 1951. Random dispersal in theoretical populations. Biometrika 38:196-218.

[70] ———. 1952. Studies in statistical ecology. I. Spatial pattern. Biometrika 39:346-62.

[71] Southwood, T. R. E. 1966. Ecological Methods, with Particular Reference to the Study of Insect Populations. Methuen, London.

[72] Spiller, D. 1948. Truncated log-normal and root-normal frequency distributions of insect populations. Nature (Lond.) 162:530.

[73] Student. 1919. An explanation of deviations from Poisson's law in practice. Biometrika 12:211-15.

[74] Sylvester, E. S., and Cox, E. L. 1961. Sequential plans for sampling aphids on sugar beets in Kern County, California. J. Econ. Ent. 54:1080-5.

[75] Taylor, L. R. 1961. Aggregation, variance and the mean. Nature (Lond.) 189:732-5.

[76] ———. 1965. A natural law for the spatial disposition of insects. Proc. 12th Intern. Congr. Ent. 1964 pp. 396-7.

[77] Thomas, M. 1949. A generalization of Poisson's binomial limit for use in ecology. Biometrika 36, 18-25.

[78] Torii, T. 1956. The stochastic approach in field population ecology with special reference to field insect populations. Japan Soc. Prom. Sci., Tokyo.

[79] Utida, S. 1943. Studies on experimental populations of the azuki bean weevil, Callosobruchus chinensis (L). VIII. Statistical analysis of the frequency distribution of the emerging weevils on beans. Mem. Coll. Agric. Kyoto Univ. 54:1-22.

[80] ———; Kono, T.; Watanabe, S.; and Yosida, T. 1952. Pattern of spatial distribution of the common cabbage butterfly, Pieris rapae in a cabbage farm. Res. Popul. Ecol. 1:49-64. (In Japanese with English summary).

[81] Watanabe, S.; Utida, S.; and Yosida, T. 1952. Dispersion of insect and change of distribution type in its process. Experimental studies on the dispersion of insects, 1. Res. Popul. Ecol. 1:94-108. (In Japanese with English summary).

[82] Waters, W. E. 1955. Sequential sampling in forest insect surveys. For. Sci. 1:68-79.

[83] ———. 1959. A quantitative measure of aggregation in insects. J. Econ. Ent. 52:1180-4.

[84] ———. 1962. The ecological significance of aggregation in forest insects. XI. Intern. Kongr. Ent. 1960, Verhandl. 2:205-10.

[85] ———, and Henson, W. R. 1959. Some sampling attributes of the negative binomial distribution with special reference to forest insects. For. Sci. 5:397-412.

[86] Williams, C. B. 1964. Patterns in the Balance of Nature and Related Problems in Quantitative Ecology. Academic Press, London and New York.

THE NEGATIVE BINOMIAL MODEL AND QUADRAT SAMPLING

J. K. ORD
Bristol University

SUMMARY

The author has suggested a new method of quadrat sampling which incorporates the two existing methods as limiting cases. For all these models, the extent of the validity of the Poisson and generalized Poisson models has been explored. For testing the assumptions underlying these models and also for distinguishing between 'true' and 'apparent' contagion, existing methods have been reviewed and certain new results added.

1. INTRODUCTION

Suppose we have a large study area and that we wish to examine a spatial distribution of points in that area (for example, a type of plant in a field). Then one approach is to select, at random, a number of small areas, or quadrats, and count the number of points in each quadrat.

Early models of the above procedure assumed that each quadrat was of equal 'size' (usually area, but not necessarily so) and was small compared to the total area; further, it was assumed that there was no interaction between quadrats. We refer to this as Model 1. More recently, the model has been extended to consider contiguous quadrats, which may be sampled in such a way as to exhaust the study area (e.g., by placing a rectangular or hexagonal grid over the entire study area). For a general discussion, see Greig Smith [9, 10]. In this second situation, Model 2, the assumption of no interaction between neighboring quadrats looks unrealistic, especially as the size of quadrat is often arbitrary or is dictated by convenience; nevertheless, fitting a generalized Poisson model to data so generated often yields a 'good fit' (e.g., see Harvey [12]).

In section 2 we study a method of quadrat sampling which includes the two usual methods discussed above as limiting cases, while in sections 3 and 4 we explore how the range of validity of the generalized Poisson model may be extended. In section 5 we review the use of sampling over two distinct time intervals to distinguish the 'true' and 'apparent' contagion hypotheses for the negative binomial distribution (NBD). This leads us to several problems concerned with testing the assumptions underlying the use of generalized Poisson models and the choice between such models, to which we turn in section 6.

2. A GENERAL METHOD OF QUADRAT SAMPLING

The 'classical' model assumes that each quadrat is small in relation to the total study area and that every quadrat is sufficiently far from every other one. The other model considered above lays a grid over the study area and samples it exhaustively. We now consider an intermediate situation which includes Models 1 and 2 as limiting cases.

First, a grid is laid over the study area defining 'sub-areas' of equal 'size.' Then, from the interior of each such sub-area (the interior is considered to avoid boundary problems) a quadrat of finite size is selected. For example, a map may be partitioned into kilometer squares, which may in turn be divided into squares with sides of 100 metres. The quadrat, then, could be visualized as one or more of these smaller squares selected at random.

If we use a constant sampling fraction, c, where

$$c = \frac{\text{size of quadrat}}{\text{size of sub-area}} \tag{2.1}$$

and the sub-area contains n points, then the probability that the quadrat will contain r such points is

$$p(r \mid n) = \binom{n}{r} c^r (1-c)^{n-r}, \quad r = 0,1,\ldots,n, \tag{2.2}$$

assuming the points to be randomly distributed within the sub-area. If the points are scattered at random over the whole study area, then the probability of there being n in a particular sub-area is

$$q_n = e^{-\lambda} \lambda^n/n! \quad , \; n = 0,1,\ldots, \tag{2.3}$$

where $E(n) = \lambda$. The observed frequency distribution for r over the different sampled areas is the generalized distribution Binomial Λ Poisson (using the notation of Gurland [11]), which is Poisson with parameter λc.

For n large, we may take c small so that (2.1) tends to the Poisson, $nc = \lambda^*$ yielding the classical model. Alternatively, allowing $c \to 1$, n finite, we arrive at the 'exhaustive sampling' model.

If this model provided no more than the above results, it would be of little value, since it does not help to distinguish the 'true' and 'apparent' contagion models (both yield the NBD, as before), but it is of value in another direction. It is now possible to draw more than one quadrat from each sub-area; we take these to be non-overlapping as there seems little value in doing otherwise (the quadrats could be of different size, if desired). Then, corresponding to (2.2), we obtain the multinomial probability for quadrat i, with parameter c_i, to contain r_i points,

$$p(r_1, r_2 \mid n) = \binom{n}{r_1, r_2} c_1^{r_1} c_2^{r_2} d^s, \qquad (2.4)$$

$$r_1, r_2 = 1,2,\ldots; \; d = 1 - c_1 - c_2, \; s = n - r_1 - r_2.$$

If n is a Poisson variate, then multinomial $\wedge$ Poisson ~ two independent Poissons (parameters λc_1, λc_2); however, if n follows the NBD or Pascal distribution with parameters k, q, multinomial $\wedge$ Pascal ~ negative multinomial; the negative multinomial being

$$p(r_1, r_2) = \binom{k + r_1 + r_2 - 1}{r_1, r_2} (qc_1)^{r_1} (qc_2)^{r_2} (1 - q)^k, \qquad (2.5)$$

yielding
$$\text{Corr}(r_1, r_2) = q\left\{\frac{c_1 c_2}{(1 - qc_1)(1 - qc_2)}\right\}^{\frac{1}{2}} . \qquad (2.6)$$

When $c_1 = c_2$ this reduces to $qc(1 - qc)^{-1}$, $c \leq \frac{1}{2}$.

This result provides us with a basis for testing whether the correlation between quadrats in a 'sub-area' is excessive, but also shows why the NBD may fit the data even when dependence between quadrats exists for the exhaustive sampling model. For, if $c_1 = c_2 = \frac{1}{2}$, there may be high correlation between certain pairs of quadrats under this model, while the marginal distribution for r in a single quadrat is the NBD. In form, however, there will be no difference between this model and the exhaustive sampling one. Indeed, the fit might even be improved because of the positive correlation which exists!

3. THE GENERALIZED POISSON MODEL

The results of this section are all well-known, but are repeated here for ease of comparison with the next section.

Assuming that the pattern of points in the study area is random, and that each quadrat is independent of any other, then the distribution of points per quadrat is Poisson. Looked at over an interval of time, the probability of there being r points at time t, $p_r(t)$, obeys the differential equation

$$p'_r(t) = \lambda p_{r-1}(t) - \lambda p_r(t) \equiv -\lambda \, \Delta p_{r-1}(t), \tag{3.1}$$

subject to $p_o(0) = 1$, where $\Delta p_{-1}(t) = p_o(t)$; $r = 0,1,\ldots.$

Thus (3.1) describes a pure birth process, and $p_r(t)$ describes a Poisson distribution with parameter λt.

If we allow 'multiple' births, then (3.1) becomes

$$p'_r(t) = -\lambda p_r(t) + \lambda \sum_{i=0}^{r} q_i \, p_{r-i}(t), \tag{3.2}$$

where q_i is the probability that the new birth creates i new points all within the quadrat ($q_o \geq 0$) and none outside the quadrat, by implication. If the $\{q_j\}$ have probability generating function (pgf)

$$G(s) = \sum_{j=0}^{\infty} q_j \, s^j,$$

then the pgf for $\{p_r(t)\}$ is

$$P(s) = \exp\,[\lambda t\{G(s) - 1\}], \tag{3.3}$$

the generalized Poisson distribution.

Thus, considered at a single point in time, our count reveals a generalized or simple Poisson process, depending on whether or not there is an element of 'clustering.' When G(s) is the pgf of the logarithmic series distribution (LSD), P(s) describes the NBD, which is indistinguishable from the gamma mixing model Poisson V Gamma ~ Pascal.

4. VALIDITY OF THE GENERALIZED POISSON MODEL

For the 'exhaustive sampling' model, a key assumption is that the numbers observed in different quadrats are mutually independent. To investigate how this assumption might be relaxed, consider the artificial situation where (N+1) quadrats are joined in a circle, the probability of the jth quadrat containing r points at time t being $p_{j,r}(t) \equiv p(j,r)$. Then a model which allows for interaction between zones is

$$p'(j,r) = -\lambda \, \Delta \, p(j,r-1) - \rho \, \Delta \, p(j-1, r-1) - \rho \, \Delta \, p(j+1, r-1), \tag{4.1}$$

for $j = 0,1,\ldots,N$, mod (N+1); ρ a parameter.

Clearly (4.1) reduces to (3.1) when $\rho = 0$. Solving (4.1) we find the pgf, $P_j(s) \equiv P(s)$ for all j, is

$$P(s) = \exp\{t(\lambda+2\rho)(s-1)\}. \tag{4.2}$$

That is, this 'dependent' model is confounded with the independent model, λ and ρ being unidentified. Alternatively, we may regard this result as showing the Poisson model to be valid in either situation. There are, of course, many forms of dependence between neighboring quadrats for which such a relation does not hold, but (4.2) does further extend the validity of the Poisson model.

We may extend this result to general quadrat 'maps' by supposing that there are k_j quadrats contiguous to the jth quadrat, forming the set

$$J \equiv J(j) = \{i(1), \ldots i(k_j)\};$$

we may then specify the model by

$$p'(j,r) = -\lambda\, \Delta\, p(j,r-1) - 2\,\rho\, k_j^{-1} \sum_{i \varepsilon j} \Delta p(i,r-1) \tag{4.3}$$

reducing to (4.1) when $k_j = 2\ \forall\ j$. Immediately we obtain

$$P_j(s) \equiv P(s) = \exp\{(\lambda+2\rho)(s-1)\}, \tag{4.4}$$

as before.

Similarly, for the multiple-births case, by comparing (3.1) and (4.3) we see from (3.2) that the pgf becomes

$$P(s) = \exp\,[t(\lambda+2\rho)\ \{G(s)-1\}], \tag{4.5}$$

so that the generalized Poisson model extends in the same way. In making these extensions we have always started from the position that $p_{j,o}(0) \equiv 1\ \forall\ j$ and observed the process at a single point in time t; that is, we have considered the single time interval (0,t).

For further discussion of why 'arrival-type' processes appear to be Poisson more often than justified by theory, see Cinclar [5].

5. OBSERVATION OVER MORE THAN ONE TIME INTERVAL

Following Bates and Neyman [4] and Bates [3], observe the process over two time intervals $(0,t_1)$, (t_1, t_1+T), $T > 0$.

Then if m points were observed in a quadrat at time m, the system of equations corresponding to (3.2) is

$$p'_r(t) = -\lambda\, p_r(t) + \lambda \sum_{i=0}^{r-m} q_i\, p_{r-i}(t)$$
$$p_m(t_1) = 1;\ r \geq m, \tag{5.1}$$

yielding for Prob(r at t_1+T | m at t_1) the pgf

$$p(s \mid t_1) = s^m \exp[\lambda T\{G(s)-1\}]. \tag{5.2}$$

That is, assuming independence between quadrats, the generalized Poisson processes for two successive non-overlapping time intervals are independent. This approach may clearly be applied to any number of time intervals $(0,t_1)$, (t_1,t_2), ... (t_{c-1},t_c).

In particular, when G(s) is the pgf of the LSD, these results enable us to distinguish the two models for the NBD, since the NBD with

$$P(s) = (1-q)^k\,(1-qs)^{-k} \tag{5.3}$$

yields for (5.2), $k \propto \lambda t_i$, q constant while the gamma mixing model with scale parameter α gives k constant, $q = t_i(\alpha+t_i)^{-1}$.

To test whether contagion was 'true' or 'apparent,' one could test $H_o{:}q_i = q$, $i = 1,\ldots,c$, against $H_1{:}q_o < q_1 < \ldots < q_c$ for the c non-overlapping time intervals. For various estimators and their asymptotic variances, see Anscombe [1].

However, when there is dependence between quadrats and m_i points are observed in the ith quadrat at time t_1, (4.3) becomes

$$p'(j,r) - -\lambda\, \Delta\, p(j,\, r-1) - 2\,\rho\, k_j^{-1} \sum_{i\varepsilon j} \Delta p(i,\, r-1) \tag{5.4}$$
$$r = m_i,\ m_i + 1,\ldots,$$

where $p(i, m_i) = 1$ at time t_1.

Unless $m_i = m$ for all i at time t_1, this system of equations no longer has (4.4) as its solution. The same argument holds for the generalized Poisson.

We may conclude, therefore, that the generalized Poisson model may 'fit the facts' even when there is dependence between quadrats, but that such dependence may be discovered by comparing results for more than one time interval. The test procedure would be as follows:

(a) fit generalized Poisson (GP) data for first period;

(b) fit GP to data for second (and subsequent) periods using the estimates derived from (a);

(c) test the resulting 'goodness of fit'. If the decision is to reject the GP model of (b), then either the parameters of the process have changed, the whole process has altered or there is dependence between quadrats;

(d) fit a GP to data for the second period, estimating the parameters from these data. If the model now fits, this implies a change in the parameter values; and

(e) if the fit in (d) is still poor, test for dependence between quadrats (see below). Finally, conclude that there is dependence between quadrats and/or that the process has altered more fundamentally.

We now turn to the mechanics of carrying out these various steps.

6. TESTS OF HYPOTHESES

In the previous section we have suggested a procedure which involves four main ingredients:

(a) choice of Poisson or generalized Poisson model;

(b) choice between GP models;

(c) estimation for the selected GP model; and

(d) check for dependence between quadrats.

We do not consider estimation problems here, but refer to Sprott [16] and Hinz and Gurland [13].

6.1 Poisson or GP model.

A variety of tests have been proposed for testing the null hypothesis that the pattern of points in the study area is random (see Greig-Smith, [10]). In particular, David and Moore [8], among others, have proposed the use of the index of dispersion or 'clumping,' $I_p = \mu_2/\mu_1'$ (μ_i denoting population moments about the mean, or about the origin if primed).

The sample index, therefore, is

$$I = s^2/\bar{x} \equiv \frac{\text{sample estimate of variance}}{\text{sample mean}}, \qquad (6.1)$$

where we take $E(s^2) = \mu_2$, a convenient but not necessary condition. Then for a sample of size n from a Poisson population with parameter λ,

$$E \equiv E(I) = \sum_{\underset{\sim}{x}} \frac{s^2}{\bar{x}} \prod_{i=1}^{n} e^{-\lambda} \lambda^{x_i}/x_i!. \tag{6.2}$$

Following Bartko et al. [2], (6.2) yields the differential equation

$$\frac{dE}{d\lambda} = n\lambda^{-2} E(s^2) - nE, \tag{6.3}$$

whence $E = 1 - ce^{-n\lambda}$, c an arbitrary constant. When $\lambda \to 0$, E(I) approaches 1 with probability one, so we impose the boundary condition $E = 1$ when $\lambda = 0$, implying $c = 0$, and thus $E = 1$ for all n, λ; that is, I is an unbiased estimator for I_p. (Alternative proofs of this result exist; see the original paper.) Bartko et al. [2] go on to show that

$$E(s^{\alpha}\, \bar{x}^{\beta}) \simeq E(s^{\alpha})\, E(\bar{x}^{\beta}),$$

implying that $(n-1)s^2/\bar{x} \sim \chi^2_{(n-1)}$. They suggest that treating I as a χ^2 variable is a valid approximation for $\lambda > 1$ or $n\lambda > 6$.

In some investigations, λ may be very small, leading to a 'lumpy' distribution for I, so that a table of extreme percentiles for low λ would be useful.

If the alternative to the Poisson is the NBD, as defined by (5.3), then the approach of (6.2), (6.3) yields

$$\frac{dE}{dq} = q^{-1}n\, E(s^2) - nkE(1-q)^{-1}, \tag{6.4}$$

from which

$$E(I) = \frac{nk}{(nk+1)p}, \text{ for } k > 0. \tag{6.5}$$

Thus, the sample index of dispersion is an almost unbiased estimator of the population index ($I_p = p^{-1}$), demonstrating that a sample value for I greater than one may be interpreted as indicating 'clumping'

Using the same approach for the binomial distribution with parameters m, p, we find

$$E(I) - \frac{mn(1-p)}{mn+1}, \tag{6.6}$$

again almost unbiased, so that $I < 1$ indicates some 'regularity' in the point pattern.

6.2 Choice of GP model.

It was shown by Anscombe [1] that to test H_o: sample drawn from a NB population, against H_1: sample drawn from a population obeying the Pólya-Aeppli (PA) distribution, the third sample moment is an asympotically fully efficient test statistic. Thus we could use the ratios $I = \mu_2/\mu_1'$, and $S = \mu_3/\mu_2$ to distinguish the distributions; these ratios have previously been considered for the classical discrete distributions in Ord [15]. Hinz and Gurland [13] use similar ratios, based on the factorial cumulants to derive more efficient estimators for GP distributions. The relationships between I and S for various GP models are:

Poisson	$S = I = 1$;
Negative Binomial (Pascal)	$S = 2I - 1$;
Pólya-Aeppli (PA)	$S = 3I/2 - (2I)^{-1}$;
Neyman type A (NTA)	$S = I + 1 - I^{-1}$;
Poisson-binomial (PB)	$S = I + (1 - p)(1 - I^{-1})$;
Poisson-Pascal (PP)	$S = I + (1 - I^{-1})/(1 - q)$;

from which Figure 1 may be constructed. In particular the Poisson-binomial lies between the NTA curve and the line S = I, while the Poisson-Pascal covers the region defined by the NTA curve and S = 2I. Between S = 2I and the NBD line only J-shaped distributions occur. The ordering of the distributions implied by Figure 1, that for given I,

$$S(PP) > S(NTA) > S(PB)$$

reflects the underlying structure of the models and provides a basis for distinguishing the models on statistical grounds.

6.3 Dependence between quadrats.

The choice of quadrat size is often rather arbitrary and it is recognized that changing the quadrat size will usually lead to different results and, in particular, may change the relationship between neighboring quadrats. If sampling is on the basis of two or more quadrats drawn from sub-areas, as in section 2, then the product moment correlation coefficient (s) can be calculated and the null hypothesis H_o: Corr = 0 (Poisson) or Corr defined by (2.6) (NBD) tested against general alternatives.

For the exhaustive sampling model we may use the statistic developed in Cliff and Ord [6], which is

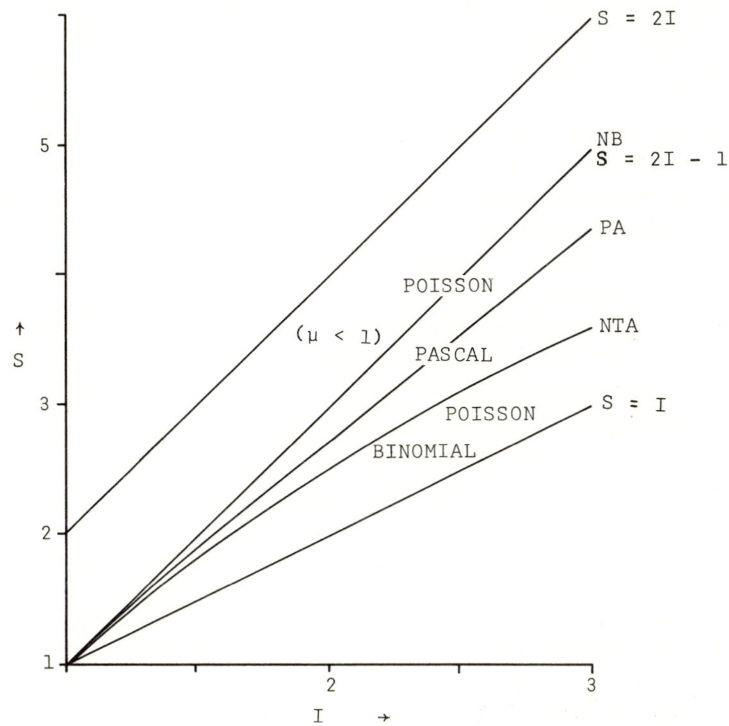

Fig. 1. I,S Region for the Generalized Poisson Distributions

$$r = \frac{n \, \Sigma_{(2)} \, w_{ij} \, z_i \, z_j}{W \sum_{i=1}^{n} z_i^2}, \tag{6.7}$$

where $z_i = x_i - \bar{x}$, deviations about the sample mean;

n = number of quadrats;

$$\Sigma_{(2)} \equiv \sum_{i=1}^{n} \sum_{\substack{j=1 \\ i \neq j}}^{n} ;$$

w_{ij} = weight attached to the link between the ith and jth quadrats; and

$W = \Sigma_{(2)} w_{ij}$.

This statistic is an extension of that proposed by Moran [14], who assumed the w_{ij} were binary weights (0 or 1). The weights may be assigned by the investigator and, in particular, may be used to search for particular 'trends' in the study area--growth of a sub-region, etc.--but for present purposes binary weights will normally be adequate. The first three moments of r and approximations for evaluating the extreme percentiles are given in Cliff and Ord [6, 7].

It is possible to invert the order of the reasoning in the previous paragraphs and to consider a large number of initial quadrats, which may be grouped in various ways. One would then select that grouping which generates the least extreme value of r. That is, either by using a clustering procedure which groups contiguous quadrats according to the criterion "minimize r" or by grouping the quadrats 1,2,3,...,k, at a time according' to some pre-specified rule. The first method will achieve a lower value for r, but this may be at the expense of a considerable loss of information or of final quadrats' being very different in size.

Alternatively, the clustering procedure could be used to identify 'homogeneous' regions within a study area. A. D. Cliff and the author are currently investigating this possibility.

REFERENCES

[1] Anscombe, F. J. 1950. Sampling theory of the negative binomial and logarithmic series distributions. Biometrika 37:358-82.

[2] Bartko, J. J.; Greenhouse, S. W.; and Patlak, C. S. 1968. On expectations of some functions of Poisson variates. Biometrics 24:97-102.

[3] Bates, G. E. 1955. Joint distributions of time intervals for the occurrence of successive accidents in a generalized Pólya scheme. Annals of Mathematical Statistics 26:705-20.

[4] ————, and Neyman, J. 1952. Contributions to the theory of accident proneness; II True or false contagion. California Publications in Statistics 1:255-75.

[5] Cinclar, E. 1968. On the superposition of m-dimensional point processes. Journal of Applied Probability 5:169-76.

[6] Cliff, A. D., and Ord, J. K. 1969a. The problem of spatial autocorrelation. In 'London Papers in Regional Science'. Ed. A. J. Scott, Pion Press, London 25-55.

[7] ————, and Ord, J. K. 1969b. The percentage points of a spatial autocorrelation Coefficient. Technical Report No. 3, Social Science Research Council Regionalisation Project, Bristol University.

[8] David, F. N., and Moore, P. G. 1954. Notes on contagious distributions in plant populations. Annals of Botany, London, New Series, 18:47-53.

[9] Greig-Smith, P. 1952. The use of random and contiguous quadrats in the study of the structure of plant communities. Annals of Botany, London, New Series, 16:293-316.

[10] ————. 1964. Quantitative Plant Ecology. Butterworth, London. Second edition.

[11] Gurland, J. 1957. Some interrelations among compound and generalized distributions. Biometrika 44:265-8.

[12] Harvey, D. W. 1966. Geographical processes and the analysis of point patterns. Transactions and Papers; Institute of British Geographers, 40:81-95.

[13] Hinz, P., and Gurland, J. 1967. Simplified techniques for estimating parameters of some generalized Poisson distributions. Biometrika 54:555-6.

[14] Moran, P. A. P. 1950. A test for the serial independence of residuals. Biometrika 37:178-81.

[15] Ord, J. K. 1967. On a system of discrete distributions. Biometrika 54:649-56.

[16] Sprott, D. A. 1965. A class of contagious distributions and maximum likelihood estimation in 'Classical and Contagious Discrete Distributions.' Ed. G. P. Patil, Pergamon Press, N. Y., 337-50.

MINIMUM CHI-SQUARE ESTIMATION FOR THE LOG-ZERO-POISSON DISTRIBUTION

A. VIJAYA RAO
Research Triangle Institute
Research Triangle Park, North Carolina

S. K. KATTI
University of Missouri

SUMMARY

Some estimators for the parameters of the log-zero-Poisson distribution are proposed. The efficiencies of twenty-seven estimators based on the minimum chi-square method are studied. It is shown that the estimator based on the sample proportions f_0/n, f_1/n, and f_2/n and the sample factorial moments $m_{(1)}$, $m_{(2)}$, and $m_{(3)}$ can be used as a substitute for the maximum likelihood estimator in certain regions of the parameter space.

1. INTRODUCTION

Several distributions have been proposed for the statistical analysis of count data. They include the Poisson, the Poisson with zeros, the negative binomial, the Neyman type A and the Poisson binomial. Katti [8] compared the skewness and kurtoksis of these and a number of other distributions developed in his unpublished Master's Thesis. It was observed that a distribution called the log-zero-Poisson (l.z.P. for brevity) has a larger range for skewness and kurtoksis than all the other distributions given in [8]. This suggests that for the purpose of describing empirical distributions, the l.z.P. distribution may be more flexible than the others.

As a check on the ability of the l.z.P. to describe empirical distributions, it was fitted to 35 empirical distributions, and the fit was compared to the fits of some other theoretical distributions. It was found to be comparable to the best among the fits by the other theoretical distributions in 32 of the 35 cases. Additional discussion on this may be found in Katti and Rao [10]. Encouraged by these empirical results, Katti and Rao [11] proposed the use of the l.z.P. as a "one-distribution summary" of the many theoretical distributions discussed above.

Basic properties of the l.z.P. distribution and the maximum likelihood estimation of the parameters are discussed in [10]. The likelihood equations have to be solved iteratively, and at every stage of the iterative procedure the probabilities have to be computed. Obviously, then, the method is very cumbersome.

The principal objective of this paper is to develop and to assess minimum chi-square methods for estimating the parameters of the l.z.P. distribution.

2. THE L.Z.P. DISTRIBUTION

The probability generating function of the l.z.P. distribution is

$$g(z) = 1 - \frac{(1-\theta)\ \ln\{q-p\exp[\lambda(z-1)]\}}{\ln\{q-p\exp(-\lambda)]}, \tag{2.1}$$

where $0 \leq \theta < 1$, $p > 0$, $\lambda > 0$, and $q = 1 + p$. On truncating the zeros in the l.z.P. distribution we get the two-parameter distribution with the probability-generating function

$$G(z) = 1 - \frac{\ln\{q-p \exp[\lambda(z-1)]\}}{\ln[q-p \exp(-\lambda)]} . \tag{2.2}$$

Note that we can get (2.2) by putting $\theta = 0$ in (2.1). For brevity the l.z.P. distribution truncated at zero will be referred to as the l.z.P.t. distribution.

In this paper we denote the probabilities in the l.z.P. distribution by $P(i)$, $i = 0,1,2,\ldots$, and those in the l.z.P.t. distribution by $P^*(i)$, $i = 1,2,\ldots$.
Clearly,

$$P(i) = (1-\theta)\ P^*(i),\ i = 1,2,\ldots, \tag{2.3}$$

and

$$P(0) = \theta. \tag{2.4}$$

3. THE INFORMATION DETERMINANT OF THE L.Z.P. DISTRIBUTION

The estimators will be compared based on their efficiency, where efficiency is defined as

$$\text{efficiency} = 1/\{\text{generalized variance x information determinant}\} \tag{3.1}$$

Therefore, in order to compute the efficiency we need the values of the information determinant. Here, we will derive the expression for the information determinant and give methods to evaluate it.

Let $I(\theta,p,\lambda)$ and $I^*(p,\lambda)$ denote the information determinants of random samples of size n from the l.z.P. distribution and the l.z.P.t. distribution, respectively. Then we have the following results:

Theorem 1.1. The information determinants $I(\theta,p,\lambda)$ and $I^*(p,\lambda)$ satisfy the relation

$$I(\theta,p,\lambda) = n\frac{(1-\theta)}{\theta}\ I^*(p,\lambda) . \tag{3.2}$$

Theorem 1.2. The information determinant $I^*(p,\lambda)$ is given by

$$I^*(p,\lambda) = n^2 \left[f_1(p,\lambda) \sum_{i=1}^{\infty} (i+1)^2 \frac{P^{*2}(i+1)}{P^*(i)} + f_2(p,\lambda)\right], \tag{3.3}$$

where

$$f_1(p,\lambda) = \frac{(1+\lambda q) \ln[q-p \exp(-\lambda)] - \lambda p}{\lambda^3 pq^2 \{\ln[q-p \exp(-\lambda)\}^2} \tag{3.4}$$

which is positive for p, $\lambda > 0$; and

$$f_2(p,\lambda) = \frac{[\exp(\lambda) -1]^2 P*^2(1)\ (1-\lambda q)}{\lambda^3 p \ln[q-p \exp(-\lambda)]} + \frac{2[\exp(\lambda) -1]P*^2(1)}{\lambda^2 p \ln[q-p\exp(-\lambda)]} - \frac{1}{\{\ln[q-p\exp(-\lambda)]\}^2}. \tag{3.5}$$

(For proof of these theorems see Rao [13]).

From these theorems we note that the information determinant of a sample from the l.z.P. distribution is the product of a simple function of θ and a second function $I*(p,\lambda)$ involving p and λ, but not θ. Hence, here we will consider only the problem of tabulating $I*(p,\lambda)$.

The difficulties in the numerical evaluation of $I*(p,\lambda)$ are due to the infinite series $s = \sum_{i=1}^{\infty} (i+1)^2 P*^2(i+1)/P*(i)$. Since $f_1(p,\lambda) > 0$, theorem 1.2 implies that the procedure of approximating S by the partial sum $S_N = \sum_{i=1}^{N} (i+1)^2 P*^2(i+1)/P*(i)$ gives only a lower bound for the information determinant. Since the value of the information determinant appears in the denominator of the expression for efficiency, this procedure will lead to an upper bound for the efficiency. To get a lower bound for the efficiency we need an upper bound for the information determinant, which is given below.

Theorem 1.3. The infinite series $S = \sum_{i=1}^{\infty} (i+1)^2 P*^2(i+1)/P*(i)$ is convergent, and the remainder after N terms of the series, $R_N = \sum_{i=N+1}^{\infty} (i+1)^2 P*^2(i+1)/P*(i)$ satisfies the inequality:

$$R_N < \frac{N+1}{\ln[q-p\exp(-\lambda)]} \left(\frac{\lambda}{\mu}\right)^{N+3} \left[1+\frac{\mu\sqrt{2}}{\sqrt{\pi(N+1)}} \exp\left(\frac{-1}{12N+13}\right)\right]^2 \Big/ \left[1-\frac{u}{\sqrt{2\pi N}} \exp\left(\frac{-1}{12N+1}\right)\right]\left[1-\left(\frac{\lambda}{\mu}\right)\frac{N+2}{N+1}\right], \tag{3.6}$$

where

$$\mu = \lambda + \ln\left(\frac{q}{p}\right), \tag{3.7}$$

and

$$N \geq \max \left\{ \frac{2\lambda-\mu}{\mu-\lambda} , \left[\frac{-1}{\ln(\lambda/\mu)} \right] \right\} . \tag{3.8}$$

For a proof of this theorem see [13].

The following procedure is followed in evaluating the information determinant: R^*_N denotes the bound for R_N. Let N be the smallest integer such that: (a) $f_1(p,\lambda)\ R^*_{N_1} < 10^{-5}$ and (b) N_1 satisfies the inequality (3.8). Then we have

$$S_{N_1} < S < S_{N_1} + R^*_{N_1} . \tag{3.9}$$

On substituting these bounds for S in the expression for the information determinant (3.3), we get upper and lower bounds which differ by less than 10^{-5}. In Table 1 we give these bounds for the information determinant of a sample of size 1 from the l.z.P.t. distribution. To obtain the bounds for the information determinants of the l.z.P. distribution, we need only to multiply the bounds corresponding to the l.z.P.t. distribution by $(1-\theta)/\theta$.

The term $(i+1)^2\ P^{*2}(i+1)/P^*(i)$ tends to zero very slowly for large values of p and λ. Consequently, the number of terms included in the partial sum S_N becomes very large as λ and p become large. In Table 2 we give the number of terms included in the partial sum, S_N, in the evaluation of the information determinant. This table will give an idea of the nature of difficulties involved in trying to extend the table of the information determinant beyond $\lambda = 5.0$ and $p = 5.0$.

4. MINIMUM CHI-SQUARE ESTIMATION

4.1 Preliminary Remarks

The minimum chi-square method is briefly as follows.

Let $x_1,\ldots,x_n$ be a random sample of size n from a distribution with density function $f(x;\theta)$, where $\theta = (\theta_1,\ldots\theta_k)'$; let $S = (s_1,\ldots,s_m)'$ be a vector of functions of the sample; and let S be a consistent asymptotically normal estimator of $\Phi = (\phi_1,\ldots,\phi_m)'$, where the ϕ's are functions of θ. Let Σ be the asymptotic covariance matrix of S. Assume that Σ is positive definite and that there exists a consistent estimator $\hat{\Sigma}$ of Σ. Let $m \geq k$ and let the Jacobian matrix

$$J = \frac{\partial \phi_i}{\partial \theta_j} \tag{4.1}$$

be of rank k.

Table 1. Upper and Lower Bounds for the Information Determinant of a Sample of Size 1 from the l.z.P.t. Distribution*

p \ λ	.5	1.0	1.5	2.0	2.5	3.0	3.5	4.0	4.5	5.0
.5	1387	2114	2570	2865	3046	3140	3170	3154	3108	3042
	1386	2113	2569	2865	3045	3139	3169	3153	3107	3041
1.0	702	932	1042	1095	1114	1110	1089	1059	1023	984
	701	932	1041	1094	1113	1109	1088	1058	1022	983
1.5	424	517	551	561	557	545	527	506	484	462
	423	516	550	560	556	544	526	505	483	461
2.0	282	324	334	333	326	315	302	288	274	260
	281	323	333	332	325	314	301	287	273	259
2.5	200	220	221	218	211	202	192	182	173	163
	200	219	220	217	210	201	191	182	172	162
3.0	149	157	156	151	146	139	132	124	117	110
	148	157	155	151	145	138	131	123	116	109
3.5	114	117	115	111	106	100	95	89	84	79
	113	117	114	110	105	99	94	88	83	78
4.0	90	91	87	84	80	75	71	67	62	59
	89	90	87	83	79	74	70	66	61	58
4.5	73	72	69	65	62	58	55	51	48	45
	72	71	68	64	61	57	54	50	47	44
5.0	60	58	55	52	49	46	43	41	38	36
	59	57	54	51	48	45	42	40	37	35

*For each value of p and λ the upper bound is given on the upper line and the lower bound on the second line. The figures given are the values of the information determinant multiplied by 10^5.

Under some regularity conditions on Φ, Barankin and Gurland [2] show that the estimator of θ obtained by minimizing the quadratic form

$$Q = (S-\Phi)' \hat{\Sigma}^{-1}(S-\Phi) \tag{4.2}$$

is asymptotically the best estimator in the class of estimators which are functions of S. This method of estimation is referred to as "the minimum chi-square method" besed on S. The estimators can be obtained by equating the derivative of Q with respect to the θ's. This leads to the system of equations

$$J' \hat{\Sigma}^{-1} (S-\phi) = 0'. \tag{4.3}$$

Table 2. Number of Terms in the Summation $\sum_{i=1}^{N} (i+1)^2 \; P*^2(i+1)/P*(i)$

p \ λ	.5	1.0	1.5	2.0	2.5	3.0	3.5	4.0	4.5	5.0
.5	14	22	29	37	44	51	59	66	73	81
1.0	18	29	39	50	61	71	82	93	104	115
1.5	21	35	49	63	77	91	105	119	133	147
2.0	24	41	58	75	92	109	127	144	169	179
2.5	28	47	67	87	108	128	148	169	189	210
3.0	31	54	77	100	123	146	169	193	217	240
3.5	34	60	86	112	138	164	191	217	244	271
4.0	37	66	95	124	153	182	211	241	271	301
4.5	40	72	103	135	167	200	232	265	298	330
5.0	43	78	112	147	182	218	253	289	324	360

The asymptotic covariance matrix of the minimum chi-square estimator of H is

$$V(\hat{\theta}) = [J' \sum{}^{-1} J]^{-1} . \qquad (4.4)$$

The properties of the minimum chi-square estimators have been studied by various authors (e.g., Barankin and Gurland [2], Chiang [3], and Ferguson [4]). This method and its variations have been applied to the estimation of the parameters of centagions distributions by several authors, including Katti and Gurland [9]. The following interesting property of the minimum chi-square estimation procedure follows trivially from the optimality of the minimum chi-squre estimators.

Theorem 4.1. Let $S_m = (s_1,...,s_m)^x$, and $S_{m+1} = (s_1,...,s_m,s_{m+1})'$. Denote by $\hat{\theta}_m$ and $\hat{\theta}_{m+1}$ the estimators of θ based on S_m and S_{m+1}, respectively. Then

$$\text{Efficiency of } \hat{\theta}_m \leq \text{efficiency of } \hat{\theta}_{m+1}.$$

This theorem enables us to form a sequence $(\hat{\theta}_m)$ of estimators with nondecreasing efficiency.

A special case of the minimum chi-square method arises when the number of statistics equals the number of parameters. The minimum chi-square estimator is the solution of the equations

$$s_i = \phi_i, \; i = 1,2,...,k.$$

The method of moments, the method of zero frequency and moments, etc., which have been used in the estimation of the parameters of contagious distributions are thus the special cases of the minimum chi-square method.

4.2 Application to the l.z.P. Distribution

Let $f_0, f_1, \ldots$ be frequencies in a sample of size n from the l.z.P. distribution. Let $s_0 = f_0/n$ and $s_i = \sum_{j=1}^{\infty} a_i(j)\, f_j/n$, $i = 1,2,\ldots,m$, where the $a_i(j)$'s are known constants and $m \geq 2$. As an example, if $a_i(j) = j(j-1)\ldots(j-i+1)$, then s_i, $i = 1,2,\ldots,m$ are the first m sample factorial moments. We will investigate some properties of the minimum chi-square estimators based on the statistics $S = (s_0, s_1, \ldots, s_m)'$.

Now, for the l.z.P. distribution, $E(s_0) = \theta$. Let $E(s_i) = \phi_i$, $i = 1,2,\ldots,m$. Then

$$\phi_i = \sum_{j=1}^{\infty} a_i(j)P(j). \tag{4.5}$$

Write

$$\Phi = (\phi_i, \ldots, \phi_m)'. \tag{4.6}$$

It can be shown that

$$\mathrm{Cov}(s_0, s_j) = \frac{-\theta\phi_j}{n}, \quad j = 1,2,\ldots,m, \tag{4.7}$$

and

$$\mathrm{Cov}(s_i s_j) = \frac{\beta_{ij} - \phi_i\phi_j}{n}, \quad i,j = 1,2,\ldots,m, \tag{4.8}$$

where

$$\beta_{ij} = \sum_{k=1}^{\infty} a_i(k)a_j(k)P(k). \tag{4.9}$$

Let

$$\phi_i^* = \sum_{k=1}^{\infty} a_i(k)P^*(k), \quad i = 1,2,\ldots,m, \tag{4.10}$$

and

$$\beta_{ij}^* = \sum_{k=1}^{\infty} a_i(k)a_j(k)P^*(k), \quad i,j = 1,2,\ldots,m, \tag{4.11}$$

where the P*(i)'s are the probabilities in the l.z.P.t. distribution.

Then

$$\beta_{ij} = (1-\theta)\ \beta^*_{ij}, i, j = 1,2,\ldots,m. \tag{4.12}$$

Write

$$\Phi^* = (\phi_1^*,\ldots,\phi_m^*)'. \tag{4.13}$$

Then

$$\Phi = (1-\theta)\Phi^* . \tag{4.14}$$

With this notation we have

$$\operatorname{cov}(s_0,s_j) = \frac{-\theta(1-\theta)}{n}\ \phi_j^* \tag{4.15}$$

and

$$\operatorname{cov}(s_i,s_j) = (1-\theta)\ \frac{\beta^*_{ij} - \phi_i^*\phi_j^*}{n} + \theta(1-\theta)\frac{\phi_i^*\phi_j^*}{n},\ i,j = 1,2,\ldots,m. \tag{4.16}$$

Let

$$\Sigma^* = \frac{\beta^*_{ij} - \phi_i^*\phi_j^*}{n} . \tag{4.17}$$

Then we can write

$$\Sigma = \begin{bmatrix} \dfrac{\theta(1-\theta)}{n} & \dfrac{-\theta(1-\theta)\Phi^{*'}}{n} \\ \dfrac{-\theta(1-\theta)\theta^*}{n} & (1-\theta)\Sigma^* + \dfrac{\theta(1-\theta)\phi^*\phi^{*'}}{n} \end{bmatrix} . \tag{4.18}$$

It can be shown that

$$\Sigma^{-1} = \begin{bmatrix} \dfrac{n}{\theta(1-\theta)} + \dfrac{\Phi^{*'}\Sigma^{*^{-1}}\Phi^*}{1-\theta} & \dfrac{\Phi^{*'}\Sigma^{*^{-1}}}{1-\theta} \\ \dfrac{\Sigma^{*^{-1}}\Phi^*}{1-\theta} & \dfrac{\Sigma^{*^{-1}}}{1-\theta} \end{bmatrix} . \tag{4.19}$$

Now, let $\hat{\theta}$ denote the minimum chi-square estimator of the parameter (θ,p,λ) based on the statistics S. Then we derive the following theorem.

<u>Theorem 4.2</u>. The efficiency of the estimator $\hat{\theta}$ is a function of p and λ and does not depend on θ.

Proof: The covariance matrix $V(\hat{\theta})$ of $\hat{\theta}$ is given by

$$[V(\hat{\theta})]^{-1} = \frac{\partial(\theta,\Phi)'}{\partial(\theta,p,\lambda)} \Sigma^{-1} \frac{\partial(\theta,\Phi)}{\partial(\theta,p,\lambda)} . \tag{4.20}$$

Now, the Jacobian matrix $\frac{\partial(\theta,\Phi)}{\partial(\theta,p,\lambda)}$ can be partitioned as

$$\frac{\partial(\theta,\Phi)}{\partial(\theta,p,\lambda)} = \begin{bmatrix} 1 & 0 \\ \Phi^* & (1-\theta)\,\frac{\partial\Phi^*}{\partial(p,\lambda)} \end{bmatrix} . \tag{4.21}$$

Hence

$$[V(\hat{\theta})] = \begin{bmatrix} \frac{n}{\theta(1-\theta)} & 0 \\ 0 & (1-\theta)\,\frac{\partial\Phi^{*\prime}}{\partial(p,\lambda)}\Sigma^{*-1}\frac{\partial\Phi^*}{\partial(p,\lambda)} \end{bmatrix} . \tag{4.22}$$

Therefore, by the definition of efficiency and by Theorem 1.1, we have

$$\text{efficiency of } \hat{\theta} = \frac{\left|\frac{\partial\Phi^{*\prime}}{\partial(p,\lambda)}\,\Sigma^{*-1}\,\frac{\partial\Phi^*}{\partial(p,\lambda)}\right|}{|I^*(p,\lambda)|} , \tag{4.23}$$

which is a function of λ and p. Hence the theorem is proved.

Now consider the statistic $T = (t_0, t_1, \ldots, t_{m-1})$, where $t_0 = s_0$ and $t_i = s_{i+1}/s_i$, $i = 1,2,\ldots,m-1$. Since S has an asymptotic $N[(\theta,\Phi)',\Sigma]$ distribution, by a well-known theorem (Rao [14], p. 322), the distribution of T is asymptotically $N[(\theta,\Psi)'\Sigma_1]$, where

$$\Psi = \left(\frac{\phi_2}{\phi_1}, \frac{\phi_3}{\phi_2}, \ldots, \frac{\phi_m}{\phi_{m-1}}\right)' = \left(\frac{\phi_2^*}{\phi_1^*}, \frac{\phi_3^*}{\phi_2^*}, \ldots \frac{\phi_m^*}{\phi_{m-1}^*}\right)' , \tag{4.24}$$

and

$$\Sigma_1 = \frac{\partial(\theta,\Psi)'}{\partial(\theta,\Phi)} \Sigma \frac{\partial(\theta,\Psi)}{\partial(\theta,\Phi)}$$

$$= \begin{bmatrix} \frac{\theta(1-\theta)}{n} & 0 \\ 0 & \frac{1}{1-\theta}\,\frac{\partial\Psi}{\partial\Phi^*}\,\Sigma^*\,\frac{\partial\Psi'}{\partial\Phi^*} \end{bmatrix} . \tag{4.25}$$

Let $\hat{\theta}^*$ denote the minimum chi-square estimator of the parameter (θ,p,λ) based on T.

<u>Theorem 4.3</u>. The efficiency of the estimator $\hat{\theta}^*$ is a function of p and λ and does not depend on θ.

Proof: The covariance matrix of $\hat{\theta}^*$ is given by

$$[V(\hat{\theta}^*)]^{-1} = \frac{\partial(\theta,\Psi)'}{\partial(\theta,p,\lambda)} \Sigma_1^{-1} \frac{\partial(\theta,\Psi)}{\partial(\theta,p,\lambda)} . \tag{4.26}$$

Now

$$\frac{\partial(\theta,\Psi)}{\partial(\theta,p,\lambda)} = \begin{bmatrix} 1 & 0 \\ 0 & \frac{\partial\Psi}{\partial(p,\lambda)} \end{bmatrix} . \tag{4.27}$$

Since the vector Ψ is a function of λ and p only, this does not involve θ. On simplifying (4.26) by using (4.25) and (4.27), we get

$$[V(\hat{\theta}^*)]^{-1} = \begin{bmatrix} \frac{n}{\theta(1-\theta)} & 0 \\ 0 & (1-\theta)\frac{\partial\Psi'}{\partial(p,\lambda)}\left[\frac{\partial\Psi}{\partial\Phi^*}\Sigma^*\frac{\partial\Psi'}{\partial\Phi^*}\right]^{-1}\frac{\partial\Psi}{\partial(p,\lambda)} \end{bmatrix} . \tag{4.28}$$

Therefore,

$$\text{efficiency of } \hat{\theta}^* = \frac{\left|\frac{\partial\Psi'}{\partial(p,\lambda)}\left[\frac{\partial\Psi}{\partial\Phi^*}\Sigma^*\frac{\partial\Psi'}{\partial\Phi^*}\right]^{-1}\frac{\partial\Psi}{\partial\Phi^*}\right|}{|I^*(p,\lambda)|} . \tag{4.29}$$

The right-hand side is a function of p and λ. This proves the theorem.

Thus the minimum chi-square estimator of (θ,p,λ) based on statistic (f_0, n, T), where T is a vector of linear function of the sample proportions $f_1/n, f_2/n, \ldots$, or a vector of ratios of such linear functions, subject to certain regularity conditions, has the property that its asymptotic efficiency is a function of p and λ only. This is to be expected, since f_0/n is the minimum variance unbiased estimator of θ.

Now, let $f_1, f_2, \ldots$ be the frequencies of counts 1,2,... in a random sample of size n from the l.z.P.t. distribution. Consider the statistic $S^* = (s_1^*, \ldots, s_m^*)'$, where

$$s_i^* = \sum_{j=1}^{\infty} a_i(j)\frac{f_j}{n}, \quad i = 1,2,\ldots,m. \tag{4.30}$$

It can be shown that

$$E(S^*) = \Phi^*, \tag{4.31}$$

and

$$\mathrm{Cov}(S^*) = \Sigma^*, \qquad (4.32)$$

where Φ^* and Σ^* are defined by (4.13) and (4.17), respectively. Let $\hat{\theta}_t$ be the minimum chi-square estimator of (p,λ) based on S^*. and recall the estimator $\hat{\theta}$.

Theorem 4.4.

$$\text{Efficiency of } \hat{\theta}_t = \text{Efficiency of } \hat{\theta}.$$

Proof:

It can be easily seen that

$$\text{efficiency of } \hat{\theta}_t = \frac{\left| \frac{\partial \Phi^{*\prime}}{\partial(p,\lambda)} \, \Sigma^{*-1} \, \frac{\partial \Phi^*}{\partial(p,\lambda)} \right|}{\left| I^*(p,\lambda) \right|} . \qquad (4.33)$$

By (4.23) this is the expression for efficiency of $\hat{\theta}$. Hence the theorem is proved.

Now, consider the statistic

$$T^* = \left(\frac{s_2^*}{s_1^*}, \ldots, \frac{s_m^*}{s_{m-1}^*} \right)' . \qquad (4.34)$$

As in the case of the statistic T, the statistic T^* is asymptotically $N(\psi, \Sigma_1^*)$, where

$$\psi = \left(\frac{\phi_2^*}{\phi_1^*}, \ldots, \frac{\phi_m^*}{\phi_{m-1}^*} \right)' ,$$

and

$$\Sigma_1^* = \frac{\partial \Psi}{\partial \Phi^*} \, \Sigma^* \, \frac{\partial \Psi'}{\partial \Phi^*} . \qquad (4.35)$$

Let $\hat{\theta}_t^*$ be the estimator of (p,λ) based on T^*; recall the estimator $\hat{\theta}^*$ of (θ,λ,p).

Theorem 4.5.

$$\text{Efficiency of } \hat{\theta}_t^* = \text{Efficiency of } \hat{\theta}^*.$$

The proof of this theorem is similar to that of theorem 4. Hence the proof is omitted.

Since f_0/n is the minimum-variance unbiased estimator of θ, we will include f_0/n in the set of statistics considered for minimum chi-square estimation. Then, by theorems proved in this section, comparison of various estimators of the three-parameter vector (θ, p, λ) is equivalent to the comparison of estimators of the two-parameter vector (p, λ) based on the same statistics computed from a random sample from the l.z.P.t. distribution. In view of this, in the next section we will compare the estimators of (p, λ) from samples from the l.z.P.t. distribution.

It should be noted that in deriving these results the particular form of the probabilities, as it occurs for the l.z.P. distribution, has not been used. Hence the results hold for more general situations.

5. EFFICIENCY OF SOME MINIMUM CHI-SQUARE ESTIMATORS

5.1 Preliminary Remarks

We will consider the minimum chi-square estimators of the vector (p, λ) based on the following statistics: the sample proportions f_1/n and f_2/n and the sample factorial moments $m_{(1)}$, $m_{(2)}$ and $m_{(3)}$. In addition to the estimator based on all the five statistics, we will study estimators based on all combinations of two, three, and four of these statistics.

One way to compare these different methods is to compare their efficiencies. In view of the difficulties associated with the evaluation of information-determinant, comparison based on efficiency will have to be only in the region $p \leq 5.0$ and $\lambda \leq 5.0$.

Now, by Theorem 4.1 the estimator based on all the five statistics has the smallest generalized variance among the estimators considered here. Hence the estimators based on subsets of the five statistics can be compared based on their relative efficiency defined as

$$\text{relative efficiency of the estimator based on a subset of statistics} = \frac{\text{generalized variance of the estimator based on all the five statistics}}{\text{generalized variance of the estimators based on the subset}}$$

The relative efficiency will be between 0 and 1 and will adequately reflect the loss in efficiency due to choosing a subset. Computation of relative efficiency involves closed algebraic expressions which are easy to evaluate on the computer in all regions of the parameter space.

In the following sections we will tabulate the efficiency of the method based on all statistics and summarize the relative efficiency of methods based on the subsets. There are 26 subsets containing two or more of the statistics. (For a listing of the subsets and tables of relative efficiency of all the methods, see Rao [13].) It may be pointed out that the efficiency of a subset can easily be calculated by the formula

efficiency as a subset

= relative efficiency of the subset x efficiency of the whole set

For the sake of brevity, we will use the phrase "method (S)" instead of "minimum chi-square method based on the statistics S."

5.2. Efficiency of the Method (f_1/n, f_2/n, $m_{(1)}$, $m_{(2)}$ and $m_{(3)}$).

Lower and upper bounds for the efficiencies for values of $p \leq 5.0$ are given in Table 3.

The efficiency of the method exceeds .94 throughout the region $\lambda \leq 2.5$ and $p \leq 5.00$ of the tabulation. For problems in which one has reason to believe that λ and p are in this region, this high efficiency might indeed justify using this simpler method in comparison with the likelihood estimators. Further justification for use of this method comes from the following considerations: (a) we have found from experience with empirical data that the likelihood function is very flat, and the solution we get empirically may be very far from the true maximum likelihood estimate; similar difficulties arise in solving likelihood equations; (b) in actual examples, sample sizes are finite, and the asymptotic efficiency given above is an approximation to the true small sample efficiency. In view of this, working too hard for small additional efficiency may not be justified.

5.3 Relative Efficiency of Methods Based on Two Statistics.

Relative efficiency of using each of the ten pairs of the statistics f_1/n, f_2/n, $m_{(1)}$, $m_{(2)}$, and $m_{(3)}$ have been tabulated. In calculating their relative efficiencies, since we do not need the value of the information determinant, the tabulation is in a larger region of the parameter space, namely $\lambda \leq 9.5$ and $p \leq 9.5$.

Table 4 gives the relative efficiency of the method (f_1,$m_{(1)}$). In the class of methods based on subsets of two statistics,

Table 3. Upper and Lower Bounds for the Efficiency of the Method

$(\frac{f_1}{n}, \frac{f_2}{n}, m_{(1)}, m_{(2)}, m_{(3)})$

p \ λ	.5	1.0	1.5	2.0	2.5	3 0	3.5	4.0	4.5	5.0
.5	1.000* .996	1.000 .994	.992 .987	.981 .974	.966 .959	.948 .942	.930 .922	.910 .901	.889 .881	.869 .859
1.0	1.000* .995	1.000* .995	.992 .987	.976 .969	.951 .944	.920 .912	.884 .874	.845 .833	.804 .791	.762 .748
1.5	1.000* .997	1.000* .995	.995 .987	.979 .969	.950 .940	.913 .901	.868 .853	.817 .801	.763 .746	.709 .689
2.0	1.000* .997	1.000* .996	.998 .987	.982 .969	.953 .938	.914 .894	.963 .842	.806 .782	.744 .718	.683 .652
2.5	1.000* .998	1.000* .995	1.000* .987	.987 .968	.958 .937	.917 .891	.865 .833	.803 .768	.737 .698	.670 .627
3.0	1.000* .997	1.000* .994	1.000* .986	.991 .968	.965 .934	.923 .887	.869 .827	.805 .758	.737 .683	.664 .608
3.5	1.000* .997	1.000* .993	1.000* .985	.999 .965	.972 .932	.933 .882	.877 .820	.812 .748	.741 .670	.669 .590
4.0	1.000* .996	1.000* .993	1.000* .984	1.000* .963	.981 .929	.942 .878	.888 .814	.824 .738	.752 .657	.676 .575
4.5	1.000* .996	1.000* .992	1.000* .982	1.000* .962	.993 .927	.957 .872	.904 .805	.839 .729	.768 .645	.692 .560
5.0	1.000* .995	1.000* .990	1.000* .980	1.000* .958	1.000* .922	.974 .866	.921 .799	.861 .717	.788 .633	.713 .546

*Upper bound for the efficiency exceeds 1.000.

Table 4. Relative Efficiency of the Method of $(f_1/n, m_{(1)})$

p \ λ	.5	1.5	2.5	3.5	4.5	5.5	6.5	7.5	8.5	9.5
.5	.886	.655	.465	.315	.202	.122	.070	.038	.020	.010
1.5	.945	.771	.594	.442	.315	.213	.136	.082	.046	.025
2.5	.960	.803	.631	.482	.385	.257	.175	.113	.068	.038
3.5	.966	.819	.649	.502	.330	.281	.199	.134	.085	.050
4.5	.970	.828	.659	.513	.393	.296	.216	.150	.098	.059
5.5	.972	.833	.666	.520	.402	.306	.227	.162	.108	.067
6.5	.974	.838	.670	.525	.403	.314	.236	.171	.117	.074
7.5	.975	.841	.674	.529	.413	.320	.243	.179	.124	.080
8.5	.976	.843	.677	.532	.416	.324	.249	.185	.130	.085
9.5	.977	.845	.679	.534	.419	.328	.253	.190	.136	.090

this method has by far the highest relative efficiency over a wide region of the parameter space.

As a summary of all the ten methods we give Table 5. The method based on $m_{(1)}$ and $m_{(2)}$ has considerable advantage over the method of (f_2 and $m_{(1)}$) in the upper triangular part of the table. For large p and λ, while the method ($m_{(1)}$ and $m_{(2)}$ is better than the method (f_2 and $m_{(1)}$), the efficiency of both of them is very poor. It is of interest to note that the remaining seven methods based on two statistics give efficiency lower than that in Table 5.

5.4 <u>Relative Efficiency of the Methods Based on Three Statistics</u>.

Relative efficiencies of all the ten methods based on these statistics have been tabulated. The method based on f_1/n, f_2/n and $m_{(1)}$ has very high relative efficiency for 2.5. The relative efficiency of this method is given in Table 6. In Table 7 is a summary of all the ten methods. The method of f_1/n, f_2/n and $m_{(1)}$ has high relative efficiency in a large part of the parameter space tabulated. The method of $m_{(1)}$, $m_{(2)}$, and $m_{(3)}$ is far better than other methods considered here for the rectangular strip at p = 0.5. There is substantial region in the parameter space where this estimator is better than the rest. The method based on f_2/n, $m_{(1)}$, and $m_{(2)}$ excels others in a small region on the parameter space of tabulation.

It should be noted that when $\lambda \leq 2.5$ the method of the subset f_1/n, f_2/n, and $m_{(1)}$ is very close to the method of the

Table 5. Highest Relative Efficiency Attained by Methods Based on Two Statistics

p \ λ	.5	1.5	2.5	3.5	4.5	5.5	6.5	7.5	8.5	9.5
.5	.957(3)*	.846(8)	.792(8)	.753(8)	.721(8)	.692(8)	.667(8)	.644(8)	.624(8)	.607(8)
1.5	.945(2)	.771(2)	.594(2)	.452(8)	.442(8)	.445(8)	.455(8)	.469(8)	.481(8)	.490(8)
2.5	.960(2)	.803(2)	.631(2)	.482(2)	.390(5)	.410(5)	.369(5)	.390(8)	.432(8)	.451(8)
3.5	.966(2)	.819(2)	.649(2)	.502(2)	.419(5)	.451(5)	.421(5)	.351(5)	.382(8)	.423(8)
4.5	.970(2)	.828(2)	.659(2)	.513(2)	.436(5)	.477(5)	.456(5)	.393(5)	.351(8)	.400(8)
5.5	.972(2)	.833(2)	.666(2)	.520(2)	.449(5)	.495(5)	.482(5)	.425(5)	.339(5)	.381(8)
6.5	.974(2)	.838(2)	.670(2)	.525(2)	.458(5)	.508(5)	.501(5)	.450(5)	.367(5)	.364(8)
7.5	.975(2)	.841(2)	.674(2)	.529(2)	.465(5)	.519(5)	.517(5)	.470(5)	.390(5)	.350(8)
8.5	.976(2)	.843(2)	.677(2)	.532(2)	.470(5)	.527(5)	.529(5)	.487(5)	.409(5)	.337(8)
9.5	.977(2)	.845(2)	.679(2)	.534(2)	.475(5)	.534(5)	.540(5)	.501(5)	.426(5)	.329(5)

*The number in parentheses is the reference number of the method: number 2 is the method $(f_1/n, m_{(1)})$; 3, the method $(f_1/n, m_{(2)})$; 5, $(f_2/n, m_{(1)})$; and 8, $(m_{(1)}, m_{(2)})$.

Table 6. Relative Efficiency of the Method (f_1/n, f_2/n, $m_{(1)}$)

p \ λ	.5	1.5	2.5	3.5	4.5	5.5	6.5	7.5	8.5	9.5
.5	.997	.949	.830	.668	.498	.346	.226	.139	.081	.045
1.5	1.000	.990	.947	.859	.733	.584	.433	.298	.191	.115
2.5	1.000	.995	.970	.912	.819	.696	.554	.410	.281	.178
3.5	1.000	.997	.980	.936	.862	.758	.630	.489	.350	.231
4.5	1.000	.998	.984	.949	.887	.797	.681	.546	.405	.275
5.5	1.000	.998	.987	.958	.904	.824	.718	.590	.450	.313
6.5	1.000	.999	.989	.963	.916	.844	.747	.624	.485	.345
7.5	1.000	.999	.990	.968	.925	.859	.769	.652	.515	.373
8.5	1.000	.999	.992	.971	.932	.871	.786	.675	.541	.398
9.5	1.000	.999	.992	.973	.937	.881	.801	.694	.563	.419

entire set f_1/n, f_2/n, $m_{(1)}$, $m_{(2)}$, $m_{(3)}$, and the loss in efficiency due to ignoring $m_{(2)}$ and $m_{(3)}$ is very small.

5.5 Relative Efficiency of Methods Based on Four Statistics.

The method (f_1/n, f_2/n, $m_{(1)}$, $m_{(2)}$) has by far the highest relative efficiency in a large region. In Table 8 the relative efficiency of this method is tabulated.

The other important estimator is that based on f_2/n, $m_{(1)}$, $m_{(2)}$, and $m_{(3)}$, which has high efficiency when p is small and when both p and λ are large. In Table 9, we give a summary of the relative efficiencies of methods based on four statistics. From this table we note that the loss in efficiency due to omitting $m_{(3)}$ is very small if $\lambda < 3.5$. Around $p = 0.5$, method (f_2/n, $m_{(1)}$, $m_{(2)}$, and $m_{(3)}$) has efficiency very close to 1. Hence the loss due to omitting is very small indeed.

We note here that, since different methods have high efficiency in different regions, methods of preliminary testing developed by Arnold [1] are appropriate. If sample size is large, any asymptotically consistent preliminary test procedure will, in fact, lead to the realization of the efficiency given in Tables 5, 7 and 9. Practical considerations will have to be used to decide which of the above methods one should use.

5.6 Relative Efficiency of the Method $\left(\frac{m_{(2)}}{m_{(1)}}, \frac{m_{(3)}}{m_{(2)}}\right)$

The estimators are obtained from the equations

Table 7. Highest Relative Efficiency Attained by Methods Based on Three Statistics

p \ λ	.5	1.5	2.5	3.5	4.5	5.5	6.5	7.5	8.5	9.5
.5	.998(20)*	.989(20)	.987(20)	.990(20)	.993(20)	.995(20)	.997(20)	.998(20)	.999(20)	.999(20)
1.5	1.000(11)	.990(11)	.947(11)	.859(11)	.760(20)	.790(20)	.831(20)	.874(20)	.913(20)	.944(20)
2.5	1.000(11)	.995(11)	.970(11)	.912(11)	.819(11)	.696(11)	.673(20)	.745(20)	.817(20)	.879(20)
3.5	1.000(11)	.997(11)	.980(11)	.936(11)	.862(11)	.758(11)	.630(11)	.649(20)	.739(20)	.823(20)
4.5	1.000(11)	.998(11)	.984(11)	.949(11)	.887(11)	.797(11)	.681(11)	.646(17)	.677(20)	.775(20)
5.5	1.000(11)	.998(11)	.987(11)	.958(11)	.904(11)	.824(11)	.718(11)	.654(17)	.635(17)	.735(20)
6.5	1.000(11)	.999(11)	.989(11)	.963(11)	.916(11)	.844(11)	.747(11)	.660(17)	.644(17)	.700(20)
7.5	1.000(11)	.999(11)	.991(11)	.968(11)	.925(11)	.859(11)	.769(11)	.665(17)	.651(17)	.670(20)
8.5	1.000(11)	.999(11)	.992(11)	.971(11)	.932(11)	.871(11)	.786(11)	.675(11)	.658(17)	.644(20)
9.5	1.000(11)	.999(11)	.992(11)	.973(11)	.937(11)	.881(11)	.801(11)	.694(11)	.663(17)	.638(17)

*The number in parentheses is the reference number of the method: number 11 is the method $(f_1/n, f_2/n, m_{(1)})$; 17, the method $(f_2/n, m_{(1)}, m_{(2)})$; and 20, $(m_{(1)}, m_{(2)}, m_{(3)})$.

Table 8. Relative Efficiency of the Method $(f_1/n, f_2/n, m_{(1)}, m_{(2)})$

p \ λ	.5	1.5	2.5	3.5	4.5	5.5	6.5	7.5	8.5	9.5
.5	.999	.984	.945	.890	.831	.774	.724	.682	.648	.621
1.5	1.000	.996	.977	.939	.882	.814	.741	.673	.617	.575
2.5	1.000	.998	.987	.960	.917	.859	.790	.719	.654	.600
3.5	1.000	.999	.991	.971	.936	.887	.825	.756	.687	.627
4.5	1.000	.999	.993	.977	.943	.905	.849	.783	.714	.650
5.5	1.000	.999	.994	.980	.955	.917	.860	.805	.736	.669
6.5	1.000	.999	.995	.983	.961	.927	.880	.821	.754	.686
7.5	1.000	.999	.996	.985	.965	.934	.891	.835	.769	.701
8.5	1.000	.999	.996	.986	.968	.940	.899	.846	.782	.713
9.5	1.000	1.000	.996	.988	.971	.944	.906	.856	.793	.724

$$\frac{\mu_{(2)}}{\mu_{(1)}} = \lambda(1+p) = \frac{m_{(2)}}{m_{(2)}} ,$$

and

$$\frac{\mu_{(3)}}{\mu_{(2)}} = \lambda(1+2p) = \frac{m_{(3)}}{m_{(2)}} .$$

The estimators are

$$\hat{\lambda} = 2\,\frac{m_{(2)}}{m_{(1)}} - \frac{m_{(3)}}{m_{(1)}} ,$$

and

$$\hat{p} = \left[\frac{m_{(3)}}{m_{(2)}} - \frac{m_{(2)}}{m_{(1)}}\right] /\hat{\lambda} \tag{5.1}$$

This estimator is very simple to calculate, but the efficiency is very low, as is seen from Table 10.

Due to the low efficiency, the use of this estimator has to be restricted to preliminary experimentation where one is looking for a quick, simple method for the sole purpose of guiding one towards more careful experimentation.

Table 9. Highest Relative Efficiency Attained by Methods Based on Four Statistics

p \ λ	.5	1.5	2.5	3.5	4.5	5.5	6.5	7.5	8.5	9.5
.5	1.000(23)*	.990(24)	.990(25)	.994(25)	.998(25)	.999(25)	.999(25)	1.000(25)	1.000(25)	1.000(25)
1.5	1.000(21)	.996(21)	.977(21)	.939(21)	.882(21)	.902(25)	.939(25)	.963(25)	.978(25)	.988(25)
2.5	1.000(21)	.998(21)	.987(21)	.960(21)	.917(21)	.859(21)	.885(25)	.925(25)	.954(25)	.973(25)
3.5	1.000(21)	.999(21)	.991(21)	.971(21)	.936(21)	.887(21)	.849(25)	.898(25)	.935(25)	.961(25)
4.5	1.000(21)	.999(21)	.993(21)	.977(21)	.947(21)	.905(21)	.849(21)	.878(25)	.920(25)	.951(25)
5.5	1.000(21)	.999(21)	.994(21)	.980(21)	.955(21)	.917(21)	.867(21)	.863(25)	.908(25)	.942(25)
6.5	1.000(21)	.999(21)	.995(21)	.983(21)	.960(21)	.927(21)	.880(21)	.851(25)	.898(25)	.935(25)
7.5	1.000(21)	.999(21)	.996(21)	.985(21)	.965(21)	.934(21)	.891(21)	.842(25)	.800(25)	.928(25)
8.5	1.000(21)	.999(21)	.996(21)	.986(21)	.968(21)	.940(21)	.899(21)	.846(21)	.882(25)	.992(25)
9.5	1.000(21)	1.000(21)	.996(21)	.988(21)	.971(21)	.944(21)	.906(21)	.856(21)	.876(25)	.917(25)

*The number in parentheses is the reference number of the method: number 21 is the method $(f_1/n, f_2/n, m_{(1)}, m_{(2)})$; 23, the method $(f_1/n, f_2/n, m_{(2)}, m_{(3)})$; 24, $(f_1/n, m_{(1)}, m_{(2)}, m_{(3)})$; and 25, $(f_2/n, m_{(1)}, m_{(2)}, m_{(3)})$.

Table 10. Relative Efficiency of the Method $\frac{m_{(2)}}{m_{(1)}}$, $\frac{m_{(3)}}{m_{(2)}}$

p \ λ	.5	1.5	2.5	3.5	4.5	5.5	6.5	7.5	8.5	9.5
.5	.515	.310	.212	.158	.127	.107	.093	.084	.077	.071
1.5	.304	.143	.085	.059	.045	.036	.031	.028	.026	.024
2.5	.198	.081	.044	.029	.022	.018	.015	.013	.013	.012
3.5	.139	.051	.027	.017	.013	.010	.009	.008	.007	.007
4.5	.101	.035	.018	.011	.008	.006	.006	.005	.005	.005
5.5	.077	.025	.013	.008	.006	.004	.004	.003	.003	.003
6.5	.060	.019	.009	.006	.004	.003	.003	.002	.002	.002
7.5	.043	.015	.007	.004	.003	.002	.002	.002	.002	.002
8.5	.039	.012	.006	.003	.002	.002	.002	.001	.001	.001
9.5	.033	.009	.005	.003	.002	.002	.001	.001	.001	.001

6. CONCLUSIONS

When the method of estimation is minimum chi-square and f_0/n is one of the statistics used, the efficiency of the estimator does not involve θ. The efficiency is equal to the efficiency of estimating λ and p using the same set of statistics from the l.z.P.t. samples, and excluding f_0/n.

The method (f_1/n, f_2/n, $m_{(1)}$, $m_{(2)}$, $m_{(3)}$) has fairly high efficiency throughout the tabulated space. In the region $\lambda \leq 2.5$, the efficiency exceeds .94; and in the region $\lambda \leq 3.5$ it exceeds .83. Therefore, in situations in which the value of λ is known to be in the region $\lambda \leq 2.5$, this method may be profitably used in preference to the more complicated maximum likelihood method. By dropping $m_{(3)}$ from the set of statistics, the loss in efficiency is very little when $\lambda \leq 2.5$. The method (f_2/n, $m_{(1)}$, $m_{(2)}$, $m_{(3)}$) has higher efficiency in the region in which the method (f_1/n, f_2/n, $m_{(1)}$, $m_{(2)}$) has relatively lower efficiency. Among the methods based on three statistics, the method (f_1/n, f_2/n, $m_{(1)}$) has high efficiency in a large region of the parameter space; the method of $m_{(1)}$, $m_{(2)}$, $m_{(3)}$ is better when p = .5 and also when λ and p are both large. Of the methods based on two statistics, the method (f_1/n, $m_{(1)}$) is better in a large part of the parameter space, but the efficiency is small.

All of the above methods have very low efficiency when λ and p are both large. This implies that the statistic $(\frac{f_1}{n}, \frac{f_2}{n}, m_{(1)}, m_{(2)}, m_{(3)})$ is a poor summary of the data when λ and p are both large. A search for statistics which reasonably summarize the data in this region is important. No attempt to find such statistics, however, has been taken up in this research.

Another interesting aspect is that these tables of efficiency seem to imply that various statistics have "ability" to improve efficiency in various regions. The statistic $m_{(3)}$ seems to improve the efficiency when λ is much larger than p. However, inclusion of $m_{(3)}$ seems to decrease the efficiency in the region $p > \lambda$. The inclusion of $\frac{f_1}{n}$ and $m_{(1)}$ leads to better efficiency in the region $p > \lambda$, and the inclusion of $\frac{f_2}{n}$ and $m_{(2)}$ improves efficiency when $p < \lambda$. For a mathematical formulation of a measure of this "ability" of various statistics so that a decision to include a statistic can be made without computing the table of efficiency, reference may be made to Petrasovits [12].

ACKNOWLEDGMENTS

This study was supported by The United States Department of Agriculture Grant 12-14-100-9183(20). The numerical computations were performed using the CDC 6400 at the Florida State University, supported by grant GJ-367 from the National Science Foundation.

REFERENCES

[1] Arnold, J. C. 1967. Estimation through preliminary tests. Unpublished Ph.D. dissertation. The Florida State University Library, Tallahassee, Florida.

[2] Barankin, E. W., and Gurland, J. 1951. On asymptotically normal, efficient estimators: I. University of California Publications in Statistics 1:89-129.

[3] Chiang, C. L. 1956. On regular best asymptotically normal estimates. Ann. Math. Stat. 27:336-51.

[4] Ferguson, T. 1958. A method of generating best asymptotically normal estimators with application to the estimation of bacterial densities. Ann. Math. Stat. 29:1046-62.

[5] Hinz, P., and Gurland, J. 1967. Simplified techniques for estimating parameters of some generalized Poisson distributions. Biometrika 54:555-66.

[6] Katti, S. K. 1958. Some families of contagious distributions. Unpublished M.S. thesis, Iowa State University, Ames, Iowa.

[7] ————. 1960. Some aspects of statistical inference for contagious distributions. Unpublished Ph.D. dissertation, Iowa State University, Ames, Iowa.

[8] ————. 1966. Interrelations among generalized distributions and their components. Biometrics 22:44-52.

[9] ————, and Gurland, J. 1962. Efficiency of certain ' methods of estimation for the negative binomial and the Neyman type A distributions. Biometrika 49:215-26.

[10] ————, and Rao, A. V. 1965. The log-zero-Poisson distribution. FSU Statistics Report M 106.

[11] ————, and Rao, A. V. 1970. The log-zero-Poisson distribution. (Accepted for publication in Biometrics)

[12] Petrasovits, Andres. 1964. A reliability function. Unpublished Master's thesis, the Florida State University Library, Tallahassee, Florida.

[13] Rao, A. V. 1969. The log-zero-Poisson distribution. Unpublished Ph.D. dissertation. The Florida State University Library, Tallahassee, Florida.

[14] Rao, C. R. 1966. Linear Statistical Inference and Its Applications. John Wiley and Sons, New York.

A CLASS OF STATISTICAL MODELS FOR MULTIPLE COUNTS

S. W. JOSHI
G. P. PATIL
The University of Texas
The Pennsylvania State University

SUMMARY

Most commonly occurring multivariate discrete probability distributions belong to the class of sum-symmetric power series distributions (SSPSD's). This paper presents properties of SSPSD's, including those related to the structure of the ranges of SSPSD's; moments; marginal, conditional, and other derived distributions and interrelations; and regression. SSPSD's are characterized among the 'regular' multivariate power series distributions. The multiple Poisson, multinomial and negative multinomial distributions are characterized among SSPSD's. Certain interesting features of minimum variance unbiased estimations are discussed. Proofs are omitted.

1. INTRODUCTION

Consider a population of individuals each having exactly one distinct character out of (s + 1) different possible characters 0,1,2,...,s. Let the proportion of individuals having the character i be π_i, so that $0 < \pi_i < 1$, $i = 0,1,2,\ldots,s$ and $\pi_0 + \pi_1 + \pi_2 + \ldots + \pi_s = 1$. Suppose a random sample of individuals is drawn from the population, sampling being with or without replacement if the population is infinite, but with replacement if it is finite. If the observation made is the s-dimensional random vector (rv) $\underset{\sim}{x} = (x_1,x_2,\ldots,x_s)$ with x_i representing the number of individuals in the sample having the character i, then, as is well known, the rv has the distributions (i) or (ii) of Section 2, depending on whether the sample of a predetermined size n is drawn or the sampling is continued until exactly k individuals with character 0 are observed. The models in (ii) with arbitrary positive k, (iii), (iv), and (v) of Section 2 arise as limits, mixtures, and/or truncations of the models in (i) and (ii). Besides the simple situations described above, these models have found applications in several other situations in different fields. (See, e.g., Bates and Neyman [1], Chatfield, Ehrenberg and Goodhardt [2], and Patil and Bildikar [7].) Patil [6] has shown that all these probability models can be nicely expressed in the form of what he has called "sum-symmetric power series distributions." In this paper we present results of investigation into structural and estimation problems associated with this class of multivariate distributions, which form a broad picture of the first author's dissertation (Joshi [3]). Proofs which could be found in the dissertation are purposely excluded from this paper so that the properties and results would appear more together and therefore could be more easily assimilated, interpreted, and used. Results of Sections 4 and 8 are proved in Patil [6].

2. NOTATION AND TERMINOLOGY

(i) Multinomial Distribution. The rv $\underset{\sim}{x} = (x_1,x_2,\ldots,x_s)$ has the s-dimensional multinomial distribution $M(\pi_1,\pi_2,\ldots,\pi_s,n)$ if its probability function (pf) is given as

$$p(\underset{\sim}{x}) = \frac{n!}{x_1!x_2!\ldots x_s!x_0!} \pi_1^{x_1}\pi_2^{x_2} \ldots \pi_s^{x_s}\pi_0^{x_0}$$

for $x_i = 0,1,2,\ldots,n$ and $i = 1,2,\ldots,s$, such that $\sum_{i=1}^{s} x_i \leq n$,

where $0 < \pi_i$, $i = 1,2,\ldots,s$, with $\sum_{i=1}^{s} \pi_i < 1$, n a positive

integer, $x_0 = n - \sum_{i=1}^{s} x_i$, and $\pi_0 = 1 - \sum_{i=1}^{s} \pi_i$.

The rv x is said to have the s-variate singular multinomial distribution $SM(\pi_1,\pi_2,\ldots,\pi_s,n)$ if its pf is given as above

$\sum_{i=1}^{s} \pi_i = 1$ for $x_i = 0,1,2,\ldots,n$, $i = 1,2,\ldots,s$, such that

$\sum_{i=1}^{s} x_i = n$.

(ii) Negative Multinomial Distribution. The rv $\underset{\sim}{x}$ has the negative multinomial distribution $NM(\pi_1,\pi_2,\ldots,\pi_s,k)$ if its pf is given as

$$p(\underset{\sim}{x}) = \frac{\Gamma(x_1+x_2+\ldots+x_s+k)}{x_1!x_2!\ldots x_s!\Gamma(k)} \pi_1^{x_1}\pi_2^{x_2} \ldots \pi_s^{x_s}\pi_0^{k}$$

for $x_i = 0,1,2,\ldots$, $i = 1,2,\ldots,s$, where $0 < \pi_i$, $i = 1,2,\ldots,s$, and $0 < k$ with $\pi_0 = 1 - \sum_{i=1}^{s} \pi_i$.

(iii) Multiple Poisson Distribution. The rv $\underset{\sim}{x}$ has the multiple Poisson distribution $MP(\lambda_1,\lambda_2,\ldots,\lambda_s)$ if its pf is given as

$$p(\underset{\sim}{x}) = e^{-(\lambda_1+\lambda_2+\ldots+\lambda_s)} \frac{\lambda_1^{x_1}\lambda_2^{x_2}\ldots\lambda_s^{x_s}}{x_1!x_2!\ldots x_s!}$$

for $x_i = 0,1,2,\ldots$, $i = 1,2,\ldots,s$, where $0 < \lambda_i$, $i = 1,2,\ldots,s$.

(iv) Origin-truncated Models. We shall adopt the notation $OTM(\pi_1,\pi_2,\ldots\pi_s,n)$, $OTNM(\pi_1,\pi_2,\ldots,\pi_s,k)$, and $OTMP(\lambda_1,\lambda_2,\ldots,\lambda_s)$ for the origin-truncated forms of $M(\pi_1,\pi_2,\ldots,\pi_s)$, $NM(\pi_1,\pi_2,\ldots,\pi_s,k)$ and $MP(\lambda_1,\lambda_2,\ldots,\lambda_s)$, respectively.

(v) Multivariate Logarithmic Distribution. The rv $\underset{\sim}{x}$ has the s-variate logarithmic distribution $ML(\pi_1,\pi_2,\ldots,\pi_s)$ if its pf is given as

$$p(\underset{\sim}{x}) = \frac{(x_1+x_2+\ldots+x_s-1)!}{x_1!x_2!\ldots x_s!} \cdot \frac{\pi_1^{x_1}\pi_2^{x_2}\ldots\pi_s^{x_s}}{[-\log(1-\pi_1-\pi_2-\ldots-\pi_s)]}$$

for $x_i = 0,1,2,\ldots,\ i = 1,2,\ldots,s$, with $\sum_{i=1}^{s} x_i > 0$, where $0 < \pi_i,\ i = 1,2,\ldots,s$, such that $\sum_{i=1}^{s} \pi_i < 1$.

(vi) Modified Logarithmic Distribution. The univariate modified logarithmic distribution with parameters δ and θ is defined by the pf

$$p(x) = \begin{cases} \delta, & x = 0 \\ \dfrac{(1-\delta)\theta^x}{-x\log(1-\theta)}, & x = 1,2,\ldots, \end{cases}$$

where $0 < \delta < 1$ and $0 < \theta < 1$.

(vii) $\underset{\sim}{e}_i$ will denote the i-th unit vector $(0,0,\ldots,0,1,0,\ldots,0)$, where 1 is in the i-th place.

3. THE CLASS OF SUM-SYMMETRIC MULTIVARIATE POWER SERIES DISTRIBUTION

The s-dimensional rv $\underset{\sim}{x}$ with the pf

$$p(\underset{\sim}{x};\underset{\sim}{\theta}) = \frac{a(\underset{\sim}{x}) \prod_{i=1}^{s} \theta_i^{x_i}}{f(\underset{\sim}{\theta})},\ \underset{\sim}{x} \in T, \tag{1}$$

is said to have the s-variate power series distribution (PSD) with range T, and the series function (sf) $f(\underset{\sim}{\theta}) = \sum a(\underset{\sim}{x}) \prod_{i=1}^{s} \theta_i^{x_i}$, where $a(\underset{\sim}{x}) > 0$ for $\underset{\sim}{x} \in T$ and $a(\underset{\sim}{x}) = 0$ otherwise. The range T is ordinarily taken to be a subset of I_s, the s-dimensional space of lattice points with non-negative integer-valued components. The s-dimensional parameter space Ω consists of the parameter vectors $\underset{\sim}{\theta} = (\theta_1,\theta_2,\ldots,\theta_s)$ for which the sf $f(\underset{\sim}{\theta})$ is positive, finite, and differentiable. Clearly, then, the parameter space Ω is the region of convergence of the power series of $f(\underset{\sim}{\theta})$. The notation we shall use for the s-variate

PSD of this section is: $\underset{\sim}{x} \sim PSD(\underset{\sim}{\theta}, f(\underset{\sim}{\theta}))$. Let us define a multivariate PSD with $f(\underset{\sim}{\theta})$ to be symmetric if $f(\theta_{i_1}, \theta_{i_2}, \ldots, \theta_{i_s}) = f(\theta_1, \theta_2, \ldots, \theta_s)$ for every permutation $(i_1, i_2, \ldots, i_s)$ of $(1, 2, \ldots, s)$, where $s \geq 2$. Further, let us call it sum-symmetric if the sf is sum-symmetric, in that $f(\underset{\sim}{\theta}) = g(\theta)$ for some $g(\cdot)$, where $\theta = \sum_{i=1}^{s} \theta_i$. Thus, for the sum-symmetric PSD we have the sf

$$f(\underset{\sim}{\theta}) = g(\theta) = \sum_{x=0}^{\infty} b(x)\theta^x$$

$$= \sum_{x=0}^{\infty} b(x) \sum_{x_1+x_2+\ldots+x_s=x} \frac{x!}{x_1!x_2!\ldots x_s!} \theta_1^{x_1}\theta_2^{x_2}\ldots\theta_s^{x_s}$$

$$= \sum \frac{(x_1+x_2+\ldots+x_s)!}{x_1!x_2!\ldots x_s!} b(x_1+x_2+\ldots+x_s)\theta_1^{x_1}\theta_2^{x_2}\ldots\theta_s^{x_s} ;$$

therefore, the pf of the s-variate sum-symmetric PSD (SSPSD) is of the form

$$p(\underset{\sim}{x};\underset{\sim}{\theta}) = \frac{\frac{(x_1+x_2+\ldots+x_s)!}{x_1!x_2!\ldots x_s!} b(x_1+x_2+\ldots+x_s)\theta_1^{x_1}\theta_2^{x_2}\ldots\theta_s^{x_s}}{g(\theta)}$$

$$= \frac{a(\underset{\sim}{x})\theta_1^{x_1}\theta_2^{x_2}\ldots\theta_s^{x_s}}{g(\theta)} \ldots ; \qquad (2)$$

where $a(\underset{\sim}{x}) = \frac{(x_1+x_2+\ldots+x_s)}{x_1!x_2!\ldots x_s!} b(x_1+x_2+\ldots+x_s)$. For this distribution we shall use the notation $\underset{\sim}{x} \sim SSPSD(\underset{\sim}{\theta}, g(\theta))$. It is interesting to note that the s-dimensional models mentioned in the previous section are all special cases of the s-variate sum-symmetric PSD. Table 1 provides the individual correspondence in brief.

4. SOME GENERAL PROPERTIES OF SSPSD's

Some of the results of this section were inspired by those of Bates and Neyman [1] for the negative multinomial distribution, a special SSPSD. Thus, if $\underset{\sim}{x} \sim SSPSD(\underset{\sim}{\theta}, g(\theta))$, then we

Table 1.

Distribution	Parameter Space	Reparametrization	Series Function
$M(\pi_1,\pi_2,\ldots,\pi_s,n)$	$0<\pi_i,\pi_1+\pi_2+\ldots+\pi_s<1$	$\theta_i = \dfrac{\pi_i}{1-\pi_1-\pi_2-\ldots-\pi_s}$	$(1+\theta_1+\theta_2+\ldots+\theta_s)^n$
$NM(\pi_1,\pi_2,\ldots,\pi_s,k)$	$0<\pi_i,\pi_1+\pi_2+\ldots+\pi_s<1$	$\theta_i = \pi_i$	$(1-\theta_1-\theta_2-\ldots-\theta_s)^{-k}$
$MP(\lambda_1,\lambda_2,\ldots,\lambda_s)$	$0 < \lambda_i$	$\theta_i = \lambda_i$	$e^{\theta_1+\theta_2+\ldots+\theta_s}$
$OTM(\pi_1,\pi_2,\ldots,\pi_s,n)$	$0<\pi_i,\pi_1+\pi_2+\ldots+\pi_s<1$	$\theta_i = \dfrac{\pi_i}{1-\pi_1-\pi_2-\ldots-\pi_s}$	$(1+\theta_1+\theta_2+\ldots+\theta_s)^n-1$
$OTNM(\pi_1,\pi_2,\ldots,\pi_s,k)$	$0<\pi_i,\pi_1+\pi_2+\ldots+\pi_s<1$	$\theta_i = \pi_i$	$(1-\theta_1-\theta_2-\ldots-\theta_s)^{-k}-1$
$OTMP(\lambda_1,\lambda_2,\ldots,\lambda_s)$	$0 < \lambda_i$	$\theta_i = \lambda_i$	$e^{\theta_1+\theta_2+\ldots+\theta_s} - 1$
$ML(\pi_1,\pi_2,\ldots,\pi_s)$	$0<\pi_i,\pi_1+\pi_2+\ldots+\pi_s<1$	$\theta_i = \pi_i$	$-\log(1-\theta_1-\theta_2-\ldots-\theta_s)$

have the following properties:

Property 4.1. $(x_{i_1}, x_{i_2}, \ldots, x_{i_s}) \sim \text{SSPSD}(\theta_{i_1}, \theta_{i_2}, \ldots, \theta_{i_s};$ $g(\theta_{i_1} + \theta_{i_2} + \ldots + \theta_{i_s}))$, where $(i_1, i_2, \ldots, i_s)$ is a permutation of $(1,2,\ldots,s)$.

Property 4.2. $(x_1, x_2, \ldots, x_r) \sim \text{SSPSD}(\theta_1, \theta_2, \ldots, \theta_r,$ $g^*(\theta_1+\theta_2+\ldots+\theta_r))$, where $g^*(\alpha) = g(\alpha + \theta_{r+1} + \theta_{r+2} + \ldots + \theta_s)$.

Property 4.3. $x_1 + x_2 + \ldots + x_s \sim \text{PSD}(\theta, g(\theta))$, where $\theta = \sum_{i=1}^{s} \theta_i$.

Property 4.4. Given $x_1 + x_2 + \ldots + x_s = x$, the conditional distribution of $\underset{\sim}{x}$ is $\text{SM}(\pi_1, \pi_2, \ldots, \pi_s, x)$, where $\pi_i = \theta_i/\theta$, $i = 1,2,\ldots,s$.

Property 4.5. $(z_1, z_2, \ldots, z_r) \sim \text{SSPSD}(\zeta_1, \zeta_2, \ldots, \zeta_r,$ $g(\zeta_1+\zeta_2+\ldots+\zeta_r))$, where $z_i = \sum_j x_i$ and $\zeta_j = \sum_j \theta_i$, with $\sum_j$ standing for the summation taken over $i \in G_j$ for $j = 1,2,\ldots,r$, such that $\bigcup_{j=1}^{r} G_j = \{1,2,\ldots,s\}$ and $G_j \cap G_{j'} = \emptyset$ for $j \neq j'$.

Property 4.6. Given $z_1, z_2, \ldots, z_r$, as defined in Property 4.5, the conditional distribution of $\underset{\sim}{x}$ is the product of r independent singular multinomial distributions, the j-th singular multinomial distribution having the parameters $\pi_{ji} = \dfrac{\theta_i}{\zeta_j}$ for $i \in G_j$ and $j = 1,2,\ldots,r$.

Property 4.7. Given $x_{r+1}, x_{r+2}, \ldots, x_s$, the conditional distribution of $(x_1, x_2, \ldots, x_r)$ is $\text{SSPSD}(\theta_1, \theta_2, \ldots, \theta_r, h(\theta_1+\theta_2+\ldots+\theta_r))$, where $h(\theta_1 + \theta_2 + \ldots + \theta_r)$ depends in form on $g(\theta_1 + \theta_2 + \ldots + \theta_s)$. Further, it depends in value on the sum $x_{r+1} + x_{r+2} + \ldots + x_s$ (and not on the individual values), but is independent of $\theta_{r+1}, \theta_{r+2}, \ldots, \theta_s$.

Property 4.8. If $\underset{\sim}{x} \sim \text{SSPSD}(\underset{\sim}{\theta}, g(\theta))$ and $\underset{\sim}{y} \sim \text{SSPSD}(\underset{\sim}{\theta}, g^*(\theta))$ and $\underset{\sim}{x}$ and $\underset{\sim}{y}$ are independent, then $\underset{\sim}{x} + \underset{\sim}{y} \sim \text{SSPSD}(\underset{\sim}{\theta}, g^{**}(\theta))$, where $g^{**}(\theta) = g(\theta)g^*(\theta)$.

Property 4.9. The class of s-variate SSPSD's consists of the power series mixtures on the parameter n of $\text{SM}(\pi_1, \pi_2, \ldots, \pi_s, n)$.

To be specific, $\underset{\sim}{x} \sim \text{SSPSD}(\underset{\sim}{\theta}, g(\theta))$ if and only if the conditional distribution of $\underset{\sim}{x}$, given $\sum_{i=1}^{s} x_i = x$, is $\text{SM}(\pi_1, \pi_2, \ldots, \pi_s, x)$, with $\pi_i = \theta_i/\theta$, $i = 1, 2, \ldots, s$, and $x \sim \text{PSD}(\theta, g(\theta))$.

All the above results, as well as some to follow, can be very conveniently proved by use of probability generating functions.

In view of Property 4.7, we emphasize that the conditional distribution is sum-symmetric. Further, it depends only on the sum $x_{r+1} + x_{r+2} + \ldots + x_s$ and not on the individual values. Also, in view of Property 4.5, the distribution of $(x_1, x_2 + x_3 + \ldots + x_s)$ is sum-symmetric. Thus, for the purpose of regression study, we need to investigate only the bivariate SSPSD's. An observation of this nature for the negative multinomial distribution was made for the first time by Bates and Neyman [1]. Later, Patil [5] observed this property to be true for a wider class of models related to Bates-Neyman's.

5. STRUCTURE OF THE RANGE OF AN SSPSD

We will call $a(\underset{\sim}{x})$ in the pf of an s-variate PSD given by (1) the coefficient function of the PSD. If the PSD of x mentioned in Property 4.9 has the pf

$$p(x, \theta) = \frac{b(x)\theta^x}{g(\theta)}, \quad x \in S, \tag{3}$$

then we have the following simple lemma useful in the proof of Properties 5.2 and 5.3.

Lemma 5.1. The coefficient functions of the SSPSD given by (2) and the corresponding univariate PSD given by (3) are related by

$$a(\underset{\sim}{x}) = \binom{x}{\underset{\sim}{x}} b(x), \tag{4}$$

where $\binom{x}{\underset{\sim}{x}} = \dfrac{x!}{x_1! x_2! \ldots x_s!}$. Further, if $x \in S$, then $\underset{\sim}{x} \in T$, for all $\underset{\sim}{x}$ such that $\sum_{i=1}^{s} x_i = x$. In particular, if $x \in S$, then

$x\underset{\sim}{e}_i \in T$ for $i = 1,2,\ldots,s$. Also, if $x \in S$, then $\{0,1,2,\ldots,x\}$ constitutes a subset of the range of every component x_i.

Property 5.2. Each of the s axes of the range of the SSPSD given by (2) carries positive probability.

Property 5.3. The set I_s is the only Cartesian product that can be the range of a nondegenerate SSPSD.

6. PROPERTIES RELATING TO MOMENTS

Properties in this section will clearly bring out the relation between the variance-covariance matrix of an SSPSD and the variance-to-mean relationship for the distribution of the component-sum.

Property 6.1. For the SSPSD given by (2) we have the following, where $x = \sum_{i=1}^{s} x_i$.

(i) $E(x_i) = \mu_1 = \frac{\theta_i}{\theta}\mu = \frac{\theta_i}{\theta} E(x)$; i.e., μ_i is proportional to θ_i.

(ii) $\mathrm{cor}(x_i,x_j) = \sigma_{ij} = \frac{\theta_i\theta_j}{\theta^2}(\sigma^2 - \mu)$, $i,j = 1,2,\ldots,s$, where $\sigma^2 = \mathrm{Var}(x)$.

(iii) $\{\mathrm{cor}(x_i,x_j)\}^2 = \rho_{ij}^2 = \left(\frac{\theta_i}{\theta_i + \lambda\theta}\right)\left(\frac{\theta_j}{\theta_j + \lambda\theta}\right)$, where $\lambda = \mu/(\sigma^2 - \mu)$.

(iv) $E(x_1^{[r_1]} x_2^{[r_2]} \ldots x_s^{[r_s]}) = \mu_{[\underset{\sim}{r}]} = \left[\prod_{i=1}^{s} (\theta_i/\theta)^{r_i}\right]\mu_{[r]}$,

where $\mu_{[r]} = E(x^{[r]})$, $x^{[r]} = x(x - 1)(x - 2)\ldots(x - r + 1)$ and $r = r_1 + r_2 + \ldots + r_s$.

Property 6.2. For an s-variate non-singular SSPSD, σ_{ii} >, =, or < μ_i if and only if σ^2>, = or < μ, respectively, and also, σ_{ii} >, =, or < μ_i if and only if σ_{jj} >, =, or < μ_j for $i,j = 1,2,\ldots,s$. Further, the correlations between pairs of components are either all positive, all negative or all zero.

Property 6.3. For an s-variate nonsingular SSPSD, the following are equivalent:

(i) For any i, $i = 1,2,\ldots,s$, $\sigma_{ii} = \mu_i$.

(ii) $\sigma^2 = \mu$.

(iii) For any pair $i,j = 1,2,\ldots,s$, $i \neq j$, $\sigma_{ij} = 0$.

(iv) $\underset{\sim}{x}$ has the multiple Poisson distribution $MP(\theta_1,\theta_2,\ldots,\theta_s)$.

(v) x has the Poisson distribution with parameter $\theta = \sum_{i=1}^{s} \theta_i$.

It should be of some interest to investigate application of Property 6.3 in testing certain relevant hypotheses.

7. MARGINAL, CONDITIONAL AND COMPONENT SUM DISTRIBUTIONS AND INTERRELATIONS

From property 6.3 we conclude that if x_1 has the Poisson distribution, then $\underset{\sim}{x}$ necessarily has the multiple Poisson distribution. More generally, given a univariate distribution with range $T_1 \subset I = \{0,1,2,\ldots\}$, we may investigate existence, uniqueness, and the form of an SSPSD of which the given distribution is a marginal distribution.

Property 7.1. There exists a nonsingular s-variate SSPSD having a given univariate distribution as the marginal distribution for one of its components if and only if the assumed univariate distribution is a power series mixture on n of the binomial distribution.

Property 7.2. An s-variate SSPSD is uniquely determined in form by its single univariate marginal distribution.

Corollary 7.3. The multinomial, the negative multinomial, the multiple Poisson, and the multivariate logarithmic distributions are the only SSPSD's which admit, respectively, the binomial, the negative binomial, the Poisson, and the modified logarithmic distributions as their marginal distributions.

Property 7.4. An s-variate SSPSD is uniquely determined by its s univariate marginal distributions.

Theorem 7.5. Suppose that the marginal distribution of an i-th component of the SSPSD given by (2) is such that $p_0^{(i)} p_2^{(i)}/[p_1^{(i)}]^2$ is independent of $\underset{\sim}{\theta}$, where $p_r^{(i)} = P[x_i = r]$.

Then $p_r^{(i)} p_{r+2}^{(i)}/[p_{r+1}^{(i)}]^2$ is independent of $\underset{\sim}{\theta}$ for $r = 0,1,2,\ldots,$ and the SSPSD is either multinomial, negative multinomial, or multiple Poisson.

Note that in view of Lemma 5.1 the ratios in Theorem 7.5 are defined.

Theorem 7.6. If the SSPSD given by (2) is such that the sf of the distribution of the sum of all the components and the sf of any univariate marginal distribution are identical in form and $\underset{\sim}{\theta}$ occurs in the sf of the latter only through its power series parameter, then the SSPSD is either multinomial, negative multinomial, or multiple Poisson.

Theorem 7.7. The SSPSD given by (2) is either the multinomial, negative multinomial, or multiple Poisson if and only if for some r, $1 \leq r \leq s - 1$, an r-variate marginal distribution is the r-variate SSPSD with the sf $g(\varphi_1 + \varphi_2 + \ldots + \varphi_r)$ such that $\underset{\sim}{\theta}$ occurs in $g(\underset{\sim}{\varphi})$ only through φ, where $\varphi = \varphi_1 + \varphi_2 + \ldots + \varphi_r$ and $\underset{\sim}{\varphi} = (\varphi_1,\varphi_2,\ldots,\varphi_r)$.

Definition: An s-variate PSD is said to be regular if its range T contains the origin and the unit vectors $\underset{\sim}{e}_i$, $i = 1,2,..,s$.

Theorem 7.8. An s-variate regular PSD of the rv $\underset{\sim}{x}$ is an s-variate SSPSD if and only if the component-sum x has a PSD.

Corollary 7.9. s-variate multinomial, negative multinomial, and multiple Poisson are the only regular s-variate PSD's admitting component-sums to have the binomial, the negative binomial and the Poisson distributions, respectively.

8. REGRESSION AND CORRELATION

Property 8.1. The regression curve for a bivariate SSPSD determines the distribution uniquely in form.

Property 8.2. If $\underset{\sim}{x} \sim \text{SSPSD}(\underset{\sim}{\theta},g(\theta))$, then the multiple correlation coefficient between x_1 and $(x_2,x_3,\ldots,x_s)$ equals the ordinary correlation coefficient between x_1 and $x_2 + x_3 + \ldots + x_s$ in numerical magnitude.

9. MVU ESTIMATION OF ARBITRARY PARAMETRIC FUNCTIONS

In this and the following sections $\underset{\sim}{y} = (y_1, y_2, \ldots, y_s)$ will denote the vector of sample sums for a random sample from an s-variate distribution of the rv $\underset{\sim}{x}$, and y will stand for the sum $\sum_{i=1}^{s} y_i$. Given $T \subset I_s$ and $\underset{\sim}{r} \in I_s$, we define $T + \{r\} = \{\underset{\sim}{t} + \underset{\sim}{r} : \underset{\sim}{t} \in T\}$, where $\underset{\sim}{t} + \underset{\sim}{r} = (t_1+r_1, t_2+r_2, \ldots, t_s+r_s)$.

Lemma 9.1. Let T and S be as defined for the SSPSD given by (2) and the univariate PSD given by (3), respectively. Let $\underset{\sim}{r} \in I_s$ and $r = \sum_{i=1}^{s} r_i$. Then $T + \{\underset{\sim}{r}\} \subset T$ if and only if $S + \{r\} \subset S$.

Theorem 9.2. Let $u(\underset{\sim}{\theta})$ be a function of the parameter vector of the SSPSD given by (2) admitting a power series expansion in the components θ_i of $\underset{\sim}{\theta}$. Then a sufficient condition for $u(\underset{\sim}{\theta})$ to be MVU-estimable is that $\theta = \sum_{i=1}^{s} \theta_i$ be MVU-estimable.

Theorem 9.3. Let $u(\underset{\sim}{\theta})$ be as in Theorem 9.2. Then a sufficient condition for $u(\underset{\sim}{\theta})$ to be MVU-estimable is that the range of the SSPSD be a Cartesian product.

10. MVU ESTIMATION OF SUMS AND PRODUCTS OF POWERS OF PARAMETRIC COMPONENTS

Theorem 10.1. For the SSPSD given by (2). $\underset{\sim}{\theta}$ is componentwise MVU-estimable if and only if $\theta = \sum_{i=1}^{s} \theta_i$ is MVU-estimable.

Theorem 10.2. For the SSPSD given by (2), $(\theta_1^{r_1}, \theta_2^{r_2}, \ldots, \theta_s^{r_s})$ is componentwise MVU-estimable if and only if $\sum_{i=1}^{s} \theta_i^{r_i}$ is MVU-estimable, where $\underset{\sim}{r} = (r_1, r_2, \ldots, r_s) \in I_s$.

Theorem 10.3. For the SSPSD given by (2), $\prod_{i=1}^{s} \theta_i^{r_i}$, with $\underset{\sim}{r} \in I_s$, is MVU-estimable on the basis of a random sample of size m if and only if θ^r, for $r = \sum r_i$ and $\theta = \sum \theta_i$, is MVU-estimable on the basis of a random sample of size m and the MVU-estimators are related by

$$\widetilde{\left(\prod_{i=1}^{s} \theta_i^{r_i}\right)} = (\widetilde{\theta^r}) \cdot \binom{y - r}{\underset{\sim}{y} - \underset{\sim}{r}} \Big/ \binom{y}{\underset{\sim}{y}} , \tag{5}$$

where ~ stands for the MVU-estimator of the parametric function under it, and $\binom{y}{\underset{\sim}{y}} = \frac{y!}{y_1!y_2!\ldots y_s!}$.

Writing $\underset{\sim}{r} = \underset{\sim}{e}_i$, $i = 1,2,\ldots,s$, (5) becomes $\tilde{\theta}_i = \tilde{\theta}(y_i/y)$; i.e.,

$$\tilde{\theta}_1 : \tilde{\theta}_2 : \ldots. : \tilde{\theta}_s = y_1 : y_2 : \ldots : y_s. \tag{6}$$

<u>Theorem 10.4</u>. The <u>pf</u> $P = P(\underset{\sim}{k};\underset{\sim}{\theta})$ of the SSPSD given by (2) is MVU-estimable for every sample size m and the MVU-estimator is given by

$$\tilde{P} = \tilde{p} \cdot \prod_{i=1}^{s} \binom{y_i}{k_i} \Big/ \binom{y}{k} , \tag{7}$$

where ~ stands for the MVU-estimator of the parametric function under it, and p is the <u>pf</u> $p(k;\theta)$ of the components-sum of the SSPSD with $k = \sum_{i=1}^{s} k_i$.

Results (5), (6), and (7) are interesting in that in each appear two factors of which the first factor is the MVU-estimator for the <u>univariate</u> distribution of the component-sum of the SSPSD and the second factor is a simple multiplying factor of binomial or multinomial coefficients. This factorization reduces the estimation problem from the multivariate to the univariate case, and tabulations for the univariate case can be convenient and useful for the multivariate case. For example, for s = 2, (5) reduces to $\widetilde{(\theta_1^{r_1}\theta_2^{r_2})} = (\widetilde{\theta^r})\ H(r_1,r_2,r;y_1,y_2,y)/\binom{r}{r_1}$, where extensive tabulations are available in print for the hypergeometric probabilities H and the binomial coefficients. Also we may observe that if θ is MVU-estimable, then $(\widetilde{\theta^r}) = \prod_{i=1}^{r-1} \tilde{\theta}(y-i;m)$, where $\tilde{\theta}(y-i;m)$ is the MUV-estimator of θ for the sample size m and the sample total y - i, implying univariate tabulation for $\tilde{\theta}$ only, would be enough.

It is interesting to note also that the second hypergeometric factor $\prod \binom{y_i}{k_i} \Big/ \binom{y}{k}$ in (7) is the MVU-estimator of the <u>pf</u> of a singular multinomial distribution at $\underset{\sim}{k}$ (Patil [4]). Also the conditional distribution of $\underset{\sim}{x}$, given x for an SSPSD, is a singular multinomial with parameters x and $\pi_i = \theta_i/\theta$, $i = 1,2,\ldots,s$. Thus we have in (7) for an SSPSD the MVU-estimator for the pf of $\underset{\sim}{x}$ = MVU-estimator for the pf of x times the MVU-estimator for the pf of the conditional distribution of $\underset{\sim}{x}$ given x. Thus the property of the pf of x is inherited by its MVU-estimator. Similar observations may be made for

$$\prod_{i=1}^{s} \theta_i^{r_i} = \theta^r \cdot \prod_{i=1}^{s} \pi_i^{r_i}.$$

11. CHARACTERIZATIONS OF AN SSPSD BY MVU-ESTIMATORS

It is not obvious that (6) should turn out to be a characterizing structure of the MVU-estimators of the class of SSPSD's among the class of multivariate PSD's. The next two theorems are designed to establish this.

Definition: Given $A \subset I_s$ and $B \subset I_s$, A is called a basis of B if m[A] = B for some positive integer m, where m[A] is defined to be the set of vectors of the form $\underset{\sim}{a}_1 + \underset{\sim}{a}_2 + \ldots + \underset{\sim}{a}_m$ with $\underset{\sim}{a}_i \in A$, $i = 1,2,\ldots,m$.

Theorem 11.1. An s-variate PSD having a basis of I_s as its range is an SSPSD if the MVU-estimators of the components of its parameter vector $\underset{\sim}{\theta}$ are proportional to the components of the sample totals vector $\underset{\sim}{y}$.

Corollary 11.2. An s-variate PSD having a basis of I_s as its range is an SSPSD if the ratios of the MVU-estimators of θ_i and θ are equal to the ratios of the corresponding sample totals y_i and y, i.e., if $\tilde{\theta}_i/\tilde{\theta} = y_i/y$, $i = 1,2,\ldots,s$.

Theorem 11.3. A bivariate PSD admitting θ_i/θ to be MVU-estimable with $\widetilde{(\theta_i/\theta)} = y_i/y$ for some i is an SSPSD.

REFERENCES

[1] Bates, G. E., and Neyman, J. 1952. Contributions to the theory of accident proneness, I. An optimistic model of the correlation between light and severe accidents. Univ. California Publ. Statist. 1:215-53.

[2] Chatfield, C.; Ehrenberg, A. S. C.; and Goodhardt, G. J. 1966. Progress on a simplified model of stationary purchasing behaviour. J. Roy. Statist. Soc. Ser. A 129:317-67.

[3] Joshi, S. W. 1969. Certain estimation and structural problems for a class of multivariate discrete distributions. Ph.D. Dissertation, Pennsylvania State University, University Park.

[4] Patil, G. P. 1965a. On multivariate generalized power series distribution and its application to the multinomial and negative multinomial. Classical and Contagious Discrete Distributions, Ed. G. P. Patil, Statistical Publishing Society, Calcutta and Pergamon Press, 183-94.

[5] Patil, G. P. 1965b. On a class of multivariate discrete distributions akin to the Bates-Neyman model in the theory of accident proneness. Sankhyā Ser A 27:259-70.

[6] Patil, G. P. 1969. On sampling with replacement from populations with multiple characters. Sankhyā Ser B 30:355-66.

[7] Patil, G. P., and Bildikar, Sheela. 1967. Multivariate logarithmic series distribution as a probability model in population and community ecology and some of its statistical properties. J. Amer. Statist. Assoc. 62:655-74.

SOME REMARKS ON TESTING OF HYPOTHESES WHEN RANDOM COUNTS ARE INVOLVED

CARL-ERIK SARNDAL
University of Umea and
University of British Columbia

SUMMARY

This paper discusses some topics in hypothesis testing with special reference to the problem of authorship decision, where an hypothesis is a statement to the effect that a certain person wrote a disputed text, or that its author is contained in a group of persons, and the data are random counts of linguistic objects. Three principles of testing are discussed: the classical (Neyman-Pearson) procedure, the likelihood test procedure, and the Bayesian procedure. Attention is focused on the case of testing a simple hypothesis H_0 as against a composite hypothesis H_1, when the latter is composed of a finite number of simple hypotheses. In this connection a class of procedures is considered, based on a generalization of the simple likelihood ratio.

1. INTRODUCTION

The problem of authorship decision for disputed texts raises some questions of a methodological nature within the realm of statistical inference theory. The data on which the decision is is to be based are assumed to consist of random counts of linguistic objects found in written texts, e.g., words. In approaching the problem, it appears appropriate to employ the apparatus of statistical hypothesis testing. By an hypothesis we mean in this context a statement to the effect that a specified person is the author of the disputed text, or that its author is contained in a specified group of persons. The difficulties encountered in applying the techniques of statistical hypotheses testing to the authorship problem are of a general nature, and consequently, although the discussion of this paper is confined to terminology peculiar to authorship decision, it will apply more generally to similar practical situations, where similar types of data are involved.

According to a commonly accepted principle of statistical inference, an hypothesis is not accepted or rejected simply on its own standing. Rather, it is accepted or rejected by virtue of its standing relative to some other hypothesis. The data evidence that supports, or fails to support, one hypothesis in comparison with a competing hypothesis may be evaluated in terms of the "factor in favor of hypothesis H_0 as against hypothesis H_1." This term has been suggested by Good ([1], p. 31). (In this paper, however, the term is used with a slightly different meaning than the one intended by Good.) For two simple hypotheses, the "factor in favor" is identical to a ratio of likelihoods.

The distributional properties of the likelihood ratios for the random counts data are rather complex. As a consequence it is found cumbersome and complicated to apply the classical approaches to hypothesis testing as outlined by the Neyman-Pearson tradition. Two more direct and more tractable rationales of hypothesis testing are discussed, namely, those of the likelihood test procedure ([2], p. 90) and the Bayesian procedure. Ultimately, the question of which procedure one wishes to pursue is, of course, a matter of philosophical inclination.

The setting of the authorship problem is briefly as follows. A disputed text is given, as well as a set of possible authors. For each possible author there may possibly exist quantifiable information in the form of a prior probability that he is in fact the author of the disputed text. In practice, the number of possible authors is always finite. Under certain generalized circumstances, it may be convenient to proceed with a model which in effect assumes infinitely many possible authors, but this case will not be treated in this paper.

The simplest problem of testing is the one where there are only two possible authors. Then only two simple hypotheses can be formulated: one person or the other must be the author of the disputed text. This situation was prevalent in the Mosteller and Wallace [3] study of the Federalist papers. If the number of possible authors is rather large, on the other hand, then many different hypotheses may be formulated. For practical reasons, it is usually difficult to take a stand on each one of them, since the testing procedure presupposes detailed knowledge of the patterns of literary style peculiar to each person encompassed by the statement of the hypothesis. Such knowledge is gained only by time-consuming scrutiny of the texts actually known to be written by these persons. Therefore, in the situation of many possible authors, the acceptance of a certain hypothesis may have to be made on provisional grounds, i.e., until one has been able to formulate a new hypothesis that is better supported by the evidential data than any one of all previously considered hypotheses.

2. FURTHER SPECIFICATION OF THE PROBLEM SETTING

The given disputed text is considered a random sample from the total literary output of the true author. In this connection, there are many problems which we shall overlook in the present work, e.g., how word rates may differ from text to text over the span of an author's life. Some of these practical problems have been discussed by Mosteller and Wallace [3].

The random counts that form the basis of the analysis are assumed to comply with a certain distributional model. The counts are defined in reference to some classification of linguistic objects, which has been found to reflect efficiently the individual variations in written texts, so that discrimination among authors is facilitated. The random counts would consist of, for example, the observed frequencies in a classification of words, where a class of words may be defined as words with a certain spelling, or as words with synonymous meaning, or as words tied to a grammatical concept like "adjective." Or the random counts may be connected with a classification by number of letters per word, or number of words per sentence. These are examples of some of the classification criteria that have been used in practical situations; for a short review of such criteria, see [7].

Each author is assumed to have his own unique parameter vector $\underset{\sim}{\theta} = (\theta_1, \ldots, \theta_k)$, where $\theta_i > 0$ $(i = 1, \ldots, k)$ and θ_i is the true rate of occurrence of the linguistic objects belonging to the i-th class of the appropriate classification, which consists of k+1 mutually exclusive and exhaustive classes. It is not necessary to specify θ_{k+1}, since

$$\theta_{k+1} = 1 - \sum_{i=1}^{k} \theta_i .$$

The true rates θ_i are assumed to have been found by a study of long excerpts of the author's known literary production. Each vector $\underset{\sim}{\theta}$ is a point in the k-dimensional parameter space

$$\Omega = \{\underset{\sim}{\theta} : \theta_i > 0 \ (i = 1, \ldots, k), \ \sum_{i=1}^{k} \theta_i < 1\}$$

Let x_i be the observed count for the i-th class in a disputed text of length ω. The vector $\underset{\sim}{x} = (x_1, \ldots, x_k)$ is an observed outcome of the random vector $\underset{\sim}{X} = (X_1, \ldots, X_k)$ with which is associated the k-dimensional sample space χ. Clearly, $x_{k+1} = \omega - \sum_{i=1}^{k} x_i$.

A simple distributional model is the k-dimensional multinomial model

$$f(\underset{\sim}{x};\underset{\sim}{\theta}) = \{\omega! / \prod_{i=1}^{k} x_i!\} \prod_{i=1}^{k} \theta_i^{x_i}; \ x_i = 0, 1, \ldots, \omega \ (i = 1, \ldots, k),$$

$$\sum_{i=1}^{k} x_i \le \omega, \tag{2.1}$$

which may be appropriate for rather extensive classes, e.g., words classified as substantives, adjectives, verbs, etc. Another simple model is the k-dimensional Poisson model

$$f(\underset{\sim}{x};\underset{\sim}{\theta}) = \prod_{i=1}^{k} \{\lambda_i^{x_i}/x_i!\} \exp(-\lambda_i);$$

$$x_i = 0,1,\ldots \quad (i = 1,\ldots,k), \tag{2.2}$$

where $\lambda_i = \omega\theta_i$ $(i = 1,\ldots,k)$, which may be appropriate for a classification where each true rate θ_i is small, e.g., in a case where each class is defined as words of the same spelling. More complicated models involving more parameters (for example, the negative binomial distribution) may sometimes be more adequate. Some aspects on the question of the goodness of fit obtained by various distributional models for word counts in particular are considered in [3].

If M is a certain set of authors, N_M will denote the number of authors contained in the set, and Ω_M will denote the set of distinct points $\underset{\sim}{\theta}$, where each $\underset{\sim}{\theta}$ is the parameter vector belonging to one of the authors in M. Thus $\Omega_M \subset \Omega$, and Ω_M contains exactly N_M distinct points.

The set of all possible authors is denoted by I. By assumption, $2 \leq N_I < \infty$. In some instances, it may be possible to assign to each author in I a prior probability that he is in fact the true author of the disputed text. Such a set of probabilities defines a discrete probability measure $P_I(\underset{\sim}{\theta})$ on Ω_I and its subsets,

$$\int_{\Omega_I} dP_I(\underset{\sim}{\theta}) = 1.$$

By a simple hypothesis is meant a statement of this type:

The person A is the author of the disputed text, where $A \in I$.

By a composite hypothesis is meant a statement of the type:

The author of the disputed text belongs to J, where $J \subset I$, and $N_J \geq 2$.

3. TWO SIMPLE HYPOTHESES

Let A_0 and A_1 be possible authors, i.e., $A_0, A_1 \in I$. The following two simple hypotheses are then to be considered,

H_0 : A_0 is the author of the disputed text,

H_1 : A_1 is the author of the disputed text.

Let $\underset{\sim}{\theta}_0 = (\theta_{01},\ldots,\theta_{0k})$ and $\underset{\sim}{\theta}_1 = (\theta_{11},\ldots,\theta_{1k})$ be the parameter vectors corresponding to A_0 and A_1, respectively. Following [1], the factor in favor of the simple hypothesis H_0 as against the simple alternative H_1 is defined as

$$F(H_0/H_1 : \underset{\sim}{x}) = f(\underset{\sim}{x};\theta_0)/f(\underset{\sim}{x};\theta_1), \tag{3.1}$$

where $\underset{\sim}{x} = (x_1,\ldots,x_k)$ denotes an observed outcome of the random vector $\underset{\sim}{X} = (X_1,\ldots,X_k)$ with which is associated the k-dimensional sample space χ. $F(H_0/H_1 : \underset{\sim}{x})$ is also called the "likelihood ratio," a more common but less expressive term. Let

$$S_c = \{\underset{\sim}{x}: F(H_0/H_1 : \underset{\sim}{x}) < c\}, \tag{3.2}$$

where c is some constant. In particular, if $\underset{\sim}{X}$ is distributed according to the Poisson model (2.2), then

$$S_c = \{\underset{\sim}{x} : \sum_{i=1}^{k} (a_i x_i - \omega d_i) < c,\ x_i \geq 0\ (i = 1,\ldots,k)\}, \tag{3.3}$$

where

$$a_i = \log \theta_{0i}/\theta_{1i} ,$$

$$d_i = \theta_{0i} - \theta_{1i} ,$$

and ω is the length of the disputed text.

The use of the factor $F(H_0/H_1 : \underset{\sim}{x})$ as the basis for a testing procedure is uncontroversial; however, the way in which the factor is put into use will differ among the three principal testing procedures that can be distinguished. Those are the classical procedure, the likelihood test procedure and the Bayesian procedure. Although the reasoning behind each of the procedures can be assumed to be well known, a short recapitulation of the main ideas may be helpful in clarifying the argument of the following section.

The classical (or Neyman-Pearson) procedure judges test performance in terms of the error probabilities

$$\alpha = P(\text{reject } H_0 | H_0 \text{ true}) \tag{3.4}$$

and

$$\beta = P(\text{reject } H_1 | H_1 \text{ true}).$$

The constant α is given, and a test satisfying (3.4) is called a test of size α. The most powerful test (i.e., the one that maximizes the power $1 - \beta$ for a fixed size α) is obtained by taking as a critical region (i.e., that subset of the sample space χ where H_0 is rejected) a region S_c defined by (3.2). The criterion for c is that it must satisfy

$$\alpha = P(\underset{\sim}{X} \varepsilon S_c | \underset{\sim}{\theta}_0)$$

for the given α. The performance of the test is measured by the size of its power

$$1 - \beta = P(\underset{\sim}{X} \varepsilon S_c | \underset{\sim}{\theta}_1).$$

Starting from a given α, the determination of c and the resulting power becomes quite cumbersome, even for relatively simple random counts models like the multinomial (2.1) and the Poisson (2.2). For example, the distribution of the linear function $\sum_{i=1}^{k} a_i X_i$ appearing in (3.3) is not a straightforward one (unless all the a_i are equal); as a result, the classical procedure is unwieldy, and heavy numerical calculation would be necessary in pursuing it.

The likelihood test procedure is based on the argument of Hacking [2]. If it happens that the observed $\underset{\sim}{x}$ is such that $\underset{\sim}{x} \varepsilon S_c$ given by (3.2) for some preassigned constant c, then the procedure rejects H_0 at the critical ratio c. This amounts to saying that H_0 is rejected because it fails to be well supported by data $\underset{\sim}{x}$ in comparison to H_1, where the relative degree of support is reflected in the magnitude of the criterion c. To the question as to how stringent the test should be, i.e., what critical value should be chosen for c, there is no general answer. It is a matter of individual judgment in much the same way as the setting of the type I error probability α in the classical procedure, or the evaluation of the posterior odds calculated in the Bayesian procedure, to which we now turn.

By the Bayesian procedure is meant the procedure of multiplying the factor $F(H_0/H_1 : \underset{\sim}{x})$ given by (3.1) by the prior odds $O(H_0/H_1)$ for H_0 as against H_1 to obtain the posterior odds $O(H_0/H_1 : \underset{\sim}{x})$ for H_0 as against H_1; i.e.,

$$O(H_0/H_1 : \underset{\sim}{x}) = F(H_0/H_1 : \underset{\sim}{x})\ O(H_0/H_1),$$

where the ratio of prior probabilities defines the prior odds,

$$O(H_0/H_1) = P_I(\underset{\sim}{\theta}_0)/P_I(\underset{\sim}{\theta}_1).$$

The magnitude of the posterior odds becomes the basis for an evaluation of the credibility of H_0 when compared to H_1. This procedure is the only one to make use of prior probabilities.

In contrast to the classical procedure, the likelihood test procedure and the Bayesian procedure are, generally speaking, straightforward in application, they do not require knowledge of the distribution of the likelihood ratio, and they are insensitive to optional stopping. The directness of the likelihood test and Bayesian procedures as contrasted to the complexity of the classical procedure will be further accentuated as we carry on to the testing of a simple hypothesis H_0 as against a composite alternative hypothesis H_1.

4. SIMPLE NULL HYPOTHESIS VERSUS A COMPOSITE HYPOTHESIS WITH A FINITE NUMBER OF ALTERNATIVES

4.1. Characterization of the Problem

Assume that we want to test a simple hypothesis H_0 against a composite alternative H_1, namely,

H_0: A_0 is the author of the disputed text,

H_1: The author of the disputed text belongs to the set J.

We assume that $J \subset I$, $A_0 \in I$, $A_0 \notin J$, where I is the set of all possible authors. Let $N_J = m$, where $2 \leq m < \infty$. The hypothesis H_1 is logically connected with the set of simple hypotheses H_{1j} $(j = 1,\ldots,m)$, where

H_{1j}: A_{1j} is the author of the disputed text.

The question to be discussed is what impact each of the simple hypotheses H_{1j} should have in constructing a critical region suitable for rejecting H_0 when it is tested against the composite alternative H_1. We may say about H_1 that the larger m is, the more <u>diffuse</u> H_1 is. Let the parameter vector be $\underset{\sim}{\theta}_0 = (\theta_{01},\ldots,\theta_{0k})$ for A_0 and $\underset{\sim}{\theta}_{1j} = (\theta_{1j1},\ldots,\theta_{1jk})$ for A_{1j} $(j = 1,\ldots,m)$. Then the parameter space pertaining to H_1 is $\Omega_J = (\underset{\sim}{\theta}_{11},\ldots,\underset{\sim}{\theta}_{1m})$.

Let

$$S_{jc} = \{\underset{\sim}{x} : f(\underset{\sim}{x};\underset{\sim}{\theta}_0)/f(\underset{\sim}{x};\underset{\sim}{\theta}_{1j}) < c\} \quad (j = 1,\ldots,m) \tag{4.1}$$

for some constant c. This S_{jc} is the critical region for rejection of H_0 when tested against the simple alternative H_{1j} at the critical ratio c. Although we are testing H_0 against the composite hypothesis H_1, the regions S_{jc} will still be important to our argument. In particular, for the Poisson model (2.2),

$$S_{jc} = \{\underset{\sim}{x} : \sum_{i=1}^{k} (a_{ji}x_i - \omega d_{ji}) < c,\ x_i \geq 0\ (i = 1,\ldots,k)\}, \quad (4.2)$$

where

$$a_{ji} = \log \theta_{0i}/\theta_{1ji},$$

$$d_{ji} = \theta_{0i} - \theta_{1ji},$$

and ω is the length of the disputed text.

Let γ_j be the power of the most powerful size α test of H_0 as against H_{1j}. This test has the critical region S_{jc} given by (4.1), where c is chosen so as to make the test of size α. Let S be an arbitrary size α test for testing H_0 as against the composite alternative H_1, and let $1 - \beta_j$ be the power of this latter test as against the simple alternative H_{1j}. If $\gamma_j = 1 - \beta_j$ $(j = 1,\ldots,m)$, then a uniformly most powerful test exists. Otherwise, $\gamma_j \geq 1 - \beta_j$ $(j = 1,\ldots,m)$ with strict inequality for at least one j, and no uniformly powerful test exists.

It is evident that a uniformly most powerful test can exist only under unrealistically special circumstances. Consider, for example, the Poisson model (2.2) for the distribution of our random counts data. The critical region for testing H_0 as against H_{1j} is S_{jc} given by (4.2). If $k > 1$, a uniformly most powerful test does not exist (and the conclusion would be the same under the multinomial model (2.1)), except under the unusual circumstances that the m hyperplanes determining the regions S_{jc} are all parallel, i.e., if

$$a_{ji} = a_i \quad (j = 1,\ldots,m),$$

which implies that, for each i, the true rate θ_{1ji} would have to be the same for each of the hypotheses H_{1j}.

In the absence of a uniformly most powerful test one may decide that it is essential to attain maximum power for <u>one</u> particular of the simple alternatives in H_1, say, H_{1j_0}. According to the Neyman-Pearson theorem, this is always possible by using $S_{j_0 c}$ determined by (4.1) as a critical region, where c is determined to make the test of size α. If we desire to detect efficiently the hypothesis "closest" to H_0, we would be motivated to choose j_0 so that H_{1j_0} is the hypothesis for which the power $1 - \beta_j$ is the least, $j = 1,\ldots,m$. This approach was proposed by Neyman and Pearson [4].

4.2. Alternative Approaches to the Problem

Let $P_J(\underset{\sim}{\theta})$ be a discrete probability measure defined on the m-point space Ω_J and its subsets. The measure $P_J(\underset{\sim}{\theta})$ may be in compliance with the prior weighing of the alternatives as induced by the prior distribution $P_I(\underset{\sim}{\theta})$ defined on Ω_I and its subsets. If this is the case, then

$$\pi_j = P_J(\underset{\sim}{\theta}_{1j}) = P_I(\underset{\sim}{\theta}_{1j})/\int_{\Omega_J} dP_I(\underset{\sim}{\theta}) \quad (j = 1,\ldots,m).$$

However, from the purely technical point of view, it is not essential to our discussion that the π_j are derived from the prior distribution, and thus we shall allow for the possibility that the π_j could be any set of positive probabilities reflecting a weighing of the hypotheses H_{1j}, such that $\sum_{j=1}^{m} \pi_j = 1$.

One way to put the set of probabilities π_j to use is seen in the following argument. Let

$$I_{S_{jc}}(\underset{\sim}{x}) = \begin{cases} 1 \text{ if } \underset{\sim}{x} \in S_{jc} \\ 0 \text{ if } \underset{\sim}{x} \notin S_{jc} \end{cases},$$

where S_{jc}, given by (4.1), is the region where H_0 is rejected when tested against H_{1j} at the critical ratio c. Let

$$I_c(\underset{\sim}{x}) = \sum_{j=1}^{m} \pi_j I_{S_{jc}}(\underset{\sim}{x}) .$$

We note that

$$I_c(\underset{\sim}{x}) = \begin{cases} 0 \text{ if and only if } x \notin S_{jc},\ j = 1,\ldots,m, \\ 1 \text{ if and only if } x \in S_{jc},\ j = 1,\ldots,m. \end{cases}$$

Thus $I_c(\underset{\sim}{x}) = 0$ and $I_c(\underset{\sim}{x}) = 1$ correspond to the cases where common sense suggests outright acceptance and rejection of H_0, respectively; the former alternative being the one where H_0 is accepted when tested against H_{1j} for each j, j = 1,...,m, at the critical ratio c, and the latter alternative being the one where H_0 is rejected when tested against H_{1j} for each j, j = 1,...,m, at the critical ratio c. Otherwise, $0 < I_c(\underset{\sim}{x}) < 1$, and then the magnitude of $I_c(\underset{\sim}{x})$ will afford a measure indicating how strongly one should interpret the evidence against H_0 when the critical ratio is c. A value of $I_c(\underset{\sim}{x})$ close to unity will be indicative of rather definite reason to consider H_0 to be refuted.

For the problem of testing against a composite hypothesis with finitely many alternatives, our main objective is, however, to discuss a continuum of approaches generalizing the likelihood ratio approach of the preceding section. First, we observe that when testing H_0 against H_1 it seems reasonable to assess jointly in some manner the data evidences pertaining to the various simple hypotheses H_{1j} connected with H_1. This joint assessment can be done in many ways, but should produce some measure which makes a sensible evaluation of the evidence of the data under the composite hypothesis H_1. Then the resulting measure should be related to the data distribution under H_0, for example, by forming a ratio. Finally, criteria have to be set according to which this ratio has to be treated in order to effectuate the test.

To this end, define the r-norm of the data distribution under H_1 as

$$\| f \|_r = \left\{ \int_{\Omega_J} [f(\underset{\sim}{x};\underset{\sim}{\theta})]^r \, dP_J(\underset{\sim}{\theta}) \right\}^{1/r}, \tag{4.3}$$

where r is any real constant, including $+\infty$ and $-\infty$, and where $P_J(\underset{\sim}{\theta})$ is a discrete probability measure defined on the m-point space Ω_J, and where $P_J(\underset{\sim}{\theta})$ has the characteristics described above. The r-norm is itself a probability distribution if r = 1. We assume that $f(\underset{\sim}{x};\underset{\sim}{\theta}_{1j}) > 0$, all $\underset{\sim}{x} \in \chi$, and j = 1,...,m, that not all $f(\underset{\sim}{x};\underset{\sim}{\theta}_{1j})$ (j = 1,...,m) are equal for a fixed $\underset{\sim}{x}$, and that $\pi_j = P_J(\underset{\sim}{\theta}_{1j}) > 0$ (j = 1,...,m).

Furthermore, we define the r-factor in favor of the simple hypothesis H_0 against the composite alternative H_1 by

$$F_r(H_0/H_1 : \underset{\sim}{x}) - f(\underset{\sim}{x};\underset{\sim}{\theta}_0)/\,\|f\|_r, \tag{4.4}$$

where $\|f\|_r$ is given by (4.3). Let

$$S_c^r = \left[\underset{\sim}{x} : F_r(H_0/H_1 : \underset{\sim}{x}) < c\right] \tag{4.5}$$

for some pair of real constants (c, r). (Our definition is different from Good's [1] definition of "factor in favor" for composite hypotheses.) Taking $F_r(H_0/H_1 : \underset{\sim}{x})$ as a basis for a procedure of testing, we obtain in effect a continuum of procedures, as r is allowed to take on any real value.

It should be noted that the ratio $F_r(H_0/H_1 : \underset{\sim}{x})$ will be left unchanged if in place of $f(\underset{\sim}{x};\underset{\sim}{\theta})$ we consider the data distribution model $f^*(\underset{\sim}{x};\underset{\sim}{\theta})$ such that

$$f^*(\underset{\sim}{x};\underset{\sim}{\theta}) = h(\underset{\sim}{x})\ f(x;\underset{\sim}{\theta}),$$

where $h(\underset{\sim}{x})$ does not depend on θ. As a consequence, likelihood test and Bayesian procedures based on $F_r(H_0/H_1 : \underset{\sim}{x})$ would, as described below, be insensitive to optional stopping.

We defer for a moment the question of how r should be chosen, and discuss first the choice of c, for a given r, when the critical region is given by (4.5).

Following a line of reasoning similar to that of the classical procedure, c would be determined in a way so as to make the test of size α; the power $1 - \beta_j$ may then be computed for the various alternatives. One is, of course, not likely to achieve maximum power against any one of the simple hypothesis H_{1j}. The computations involved would be quite cumbersome, for example, under the Poisson model (2.2), but numerical studies could be performed.

In a generalization of the likelihood test procedure, originally proposed by Hacking for the case of r = ∞ (see below), one would settle for some constant c to reflect the stringency of the test; the smaller c, the stronger must be the evidence contrary to H_0 before H_0 is rejected.

In a generalization of the Bayesian procedure, finally, the factor $F_r(H_0/H_1 : \underset{\sim}{x})$ may be regarded as an expression of data evidence which, after multiplication by the prior odds

$$O(H_0/H_1) = P_I(\underset{\sim}{\theta}_1)/\int_{\Omega_J} dP_I(\underset{\sim}{\theta})\ ,$$

gives the posterior odds

$$O(H_0/H_1 : \underset{\sim}{x}) = F_r(H_0/H_1 : \underset{\sim}{x})\ O(H_0/H_1).$$

These may form the basis for judging the credibility of H_0 in comparison to H_1.

Turning now to the question of how to choose an appropriate value of r for a fixed c, we prove first the following three propositions about the critical region S_c^r defined by (4.5), namely,

$$(1) \quad S_c^{r_0} \subset S_c^{r_1} \quad \text{if } r_0 < r_1 ;$$

$$(2) \quad S_c^{\infty} = \bigcup_{j=1}^{m} S_{jc} ;$$

$$S_c^{-\infty} \quad \bigcap_{j=1}^{m} S_{jc} ;$$

$$\text{and} \quad (3) \quad \bigcap_{j=1}^{m} S_{jc} \subset S_c^r \subset \bigcup_{j=1}^{m} S_{jc},$$

where S_{jc} is given by (4.1), and $-\infty < r < \infty$.

Proposition (3) is an immediate consequence of (1) and (2), which we set out to prove.

Consider the function

$$\phi(r) = \|f\|_r = \left(\sum_{j=1}^{m} f_j^r \pi_j\right)^{1/r},$$

where $\pi_j > 0$ $(j = 1,\ldots,m)$, $f_j = f(\underset{\sim}{x};\underset{\sim}{\theta}_{1j}) > 0$ $(j = 1,\ldots,m)$, for $\underset{\sim}{x} \in \chi$, and not all f_j equal. We have $\phi(r) > 0$, all r, and

$$\frac{d \log \phi(r)}{dr} = r^{-2} \left(\sum_{j=1}^{m} a_j\right)^{-1} \sum_{j=1}^{m} a_j \log \frac{a_j}{b_j}, \qquad (4.6)$$

where

$$a_j = f_j^r \pi_j ,$$

$$b_j = \pi_j \sum_{i=1}^{m} f_i^r \pi_i ,$$

and hence $\sum_{j=1}^{m} a_j = \sum_{j=1}^{m} b_j > 0$. The last sum of (4.6) is positive by an elementary inequality of information theory (see, for example, Rao [6], p. 47). Hence $\phi(r)$ is increasing monotonically as r increases. It follows that

$$F_{r_0}(H_0/H_1 : \underset{\sim}{x}) > F_{r_1}(H_0/H_1 : \underset{\sim}{x}),$$

if $r_0 < r_1$, for all $\underset{\sim}{x} \in \chi$. Thus,

$$F_{r_1}(H_0/H_1 : \underset{\sim}{x}) < c, \text{ for } \underset{\sim}{x} \in S_c^{r_0},$$

and proposition (1) follows.

In proof of the first statement of proposition (2), we note that

$$\lim_{r\to\infty} ||f||_r = \max_{j=1,..,m.} f(\underset{\sim}{x};\underset{\sim}{\theta}_{1j})$$

Hence

$$F_\infty(H_0/H_1 : \underset{\sim}{x}) = f(\underset{\sim}{x};\underset{\sim}{\theta}_0)/\max.\ f(\underset{\sim}{x};\underset{\sim}{\theta}_{1j}). \tag{4.7}$$

It follows that $\underset{\sim}{x} \in S_c$ if and only if $\underset{\sim}{x} \in S_{jc}$ for <u>some</u> j; j = 1,...,m. The second statement of proposition (2) follows similarly by observing that

$$F_{-\infty}(H_0/H_1 : \underset{\sim}{x}) = f(\underset{\sim}{x};\underset{\sim}{\theta}_0)/\min_{j=1,...,m} f(\underset{\sim}{x};\underset{\sim}{\theta}_{1j}).$$

Propositions (1) and (2) tell us that the value of r will, for a fixed c and a given measure P_J, affect the stringency of the test through the structuring of the critical region. If, on the one extreme, $r = \infty$, then we reject H_0 at the critical ratio c as soon as there exists <u>some</u> H_{1j} in comparison with which H_0 fails to be supported by data. If, on the other extreme, $r = -\infty$, then we are willing to reject H_0 at the critical ratio c only if H_0 fails to be supported by data for <u>every one</u> of the H_{1j}. The pair (r,c) therefore jointly determines the stringency of the test; decreasing r and/or c will make the test more stringent, i.e., less prone to reject H_0.

In addition, the structure of the critical region S_c^r therefore requires specification of the triplet (r,c,P_J). It should be noted, however, that if $r = \pm\infty$, S_c^r becomes entirely independent of P_J.

If $r = \infty$, the test based on $F_\infty(H_0/H_1 : \underset{\sim}{x})$ given by (4.7) is essentially the procedure suggested by Neyman and Pearson [47], although their ratio is limited upwards by unity through including the null hypothesis in the denominator. Furthermore, the case of $r = \infty$ corresponds to the likelihood test procedure outlined in [2], p. 90. Some problems of the Bayesian procedure in this connection are discussed in [1], p. 32.

As we have observed earlier, the r-factor (4.4) is a ratio of probability distributions if $r = 1$,

$$F_1(H_0/H_1 : \underset{\sim}{x}) = f(\underset{\sim}{x};\underset{\sim}{\theta}_0)/\sum_{j=1}^{m} \pi_j\, f(\underset{\sim}{x};\underset{\sim}{\theta}_{1j}). \tag{4.8}$$

Hence, if the use of some probability measure $P_J(\underset{\sim}{\theta})$ can be accepted as a means of weighing the alternatives, the Neyman-Pearson theorem produces the most powerful test. This implies that if the critical region S_c^1 is determined by (4.5) for $r = 1$, and c is suited to a fixed size α, then the power

$$\int_{S_c^1} \sum_{j=1}^{m} \pi_j\, f(\underset{\sim}{x};\underset{\sim}{\theta}_{1j})\, d\underset{\sim}{x}$$

is a maximum (see [5], p. 313). To the Bayesian point of view, the procedure based on $r = 1$ may be the one having the most obvious appeal; the denominator of (4.8) representing the data distribution under H_1.

It is worth observing that the procedures based on $r = 0$ (defined as the limit when $r \to 0$) and $r = -1$ correspond to the geometric and the harmonic means, respectively, of the qualities $f(\underset{\sim}{x};\underset{\sim}{\theta}_{1j})$, so that

$$F_0(H_0/H_1 : \underset{\sim}{x}) = f(\underset{\sim}{x};\underset{\sim}{\theta}_0)/\prod_{j=1}^{m} \{f(\underset{\sim}{x};\underset{\sim}{\theta}_{1j})\}^{\pi_j} \quad (r = 0)$$

and

$$F_{-1}(H_0/H_1 : \underset{\sim}{x}) = f(\underset{\sim}{x};\underset{\sim}{\theta}_0) \sum_{j=1}^{m} \{\pi_j/f(\underset{\sim}{x};\underset{\sim}{\theta}_{1j})\} \quad (r = -1).$$

It follows from proposition (3) that if the observed outcome $\underset{\sim}{x}$ is such that $\underset{\sim}{x} \in S_{jc}$ for <u>every</u> j, $j = 1,\ldots,m$, so that H_0 is rejected for each of the simple alternatives H_{1j} at the critical ratio c, then the test based on the critical region S_c^r given by (4.5) will reject H_0 regardless of the value of r and of the probability measure $P_J(\underset{\sim}{\theta})$ used for weighing the alternatives. The policy of rejecting H_0 under any circumstances (i.e., any r and any $P_J(\underset{\sim}{\theta})$) when H_0 is in fact rejected in comparison to every one of the H_{1j} is, of course, only common sense.

The values of r represent different attitudes towards the manner in which a simple hypothesis H_0 should properly be tested against a composite alternative. As with several other matters is the area of statistical hypothesis testing, no conclusive answer can be given as to what value of r one should actually apply. A resolution of the question has to be sought in one's philosophical attitude towards the question of a proper structuring of a critical region for this problem.

REFERENCES

[1] Good, I. J. 1965. The Estimation of Probabilities. MIT Press, Cambridge, Mass.

[2] Hacking, I. 1965. Logic of Statistical Inference. Cambridge University Press, Cambridge.

[3] Mosteller, F. and Wallace, D. L. 1963. Inference and Disputed Authorship: The Federalist. Addison-Wesley, Reading, Mass.

[4] Neyman, J., and Pearson, E. S. 1933. The Testing of Statistical Hypotheses in Relation to Probabilities a priori. Proc. Cam. Phil. Soc., 29:492-510.

[5] Rao, C. R. 1952. Advanced Statistical Methods in Biometric Research. John Wiley, New York.

[6] ———. 1965. Linear Statistical Inference and Its Applications. John Wiley, New York.

[7] Särndal, C. E. 1967. On Deciding Cases of Disputed Authorship. Appl. Stat., 16:251-68.

GOODNESS OF FIT OF A COMPOUND MULTINOMIAL DISTRIBUTION AND ALLIED PROBLEMS

SUJIT KUMAR MITRA
Indian Statistical Institute
Calcutta, India

SUMMARY

This paper deals with the problem of testing goodness of fit of a multinomial distribution when the total number of possible observations (the sample size n) is an unobserved random variable with a known distribution. The data for analysis are the observed frequencies $O_1, O_2, \ldots, O_k$ in k classes of the multinomial distribution with the corresponding class probabilities $\pi_1, \pi_2, \ldots, \pi_k$ adding up to a number of strictly less than 1. Given a specified value of the random variable n, the conditional distribution of $O_1, O_2, \ldots, O_k$, $n - \Sigma O_i$, is positive multinomial. Marginally $O_1, O_2, \ldots, O_k$ follow a compound multinomial distribution. Consider a hypothesis H_0 which numerically determines the class probabilities π_i and, accordingly, the vector of expected frequencies $E_1, E_2, \ldots, E_k$. It is shown that under certain conditions the hypothesis could be tested by a suitably constructed goodness-of-fit chi-square statistic, based only on the observables $O_1, O_2, \ldots, O_k$. The case of a composite hypothesis which determines the class probabilities only as known functions of certain unknown parameters is also considered.

1. INTRODUCTION

This paper deals with the problem of testing goodness-of-fit of a multinomial distribution for which the total number of possible observations (the sample size n) is an unobserved random variable with a known distribution. The data for analysis are the observed frequencies $O_1, O_2, \ldots, O_k$ in k classes of the multinomial distribution with the corresponding class probabilities $\pi_1, \pi_2, \ldots, \pi_k$ adding up to a number strictly less than 1. Given a specified value for the random variable n, the conditional distribution of $O_1, O_2, \ldots, O_k$, $n-\Sigma O_i$, is positive multinomial. Marginally $O_1, O_2, \ldots, O_k$ follow a compound multinomial distribution. Consider now a hypothesis H_0 which numerically determines the class probabilities π_i and, accordingly, the vector of expected frequencies $E_1, E_2, \ldots, E_k$. It is shown that under certain conditions the hypothesis could be tested by a suitably constructed goodness-of-fit chi-square statistic, based only on the observables $O_1, O_2, \ldots, O_k$ (Section 3). The case of a composite hypothesis which determines the class probabilities only as known functions of certain unknown parameters is dealt with in Section 4; Section 2 gives some preliminary mathematical results.

The motivation for writing this paper is twofold. First, as far as the author knows, the problem has not received attention in previous literature except in papers by Steyn [9, 10] and Sibuya et al [8]. These authors deal with the negative multinomial distribution, which incidentally would be viewed as a multinomial distribution compounded with a negative binomial distribution for n. Second, the simplicity associated with Pearson's form of goodness-of-fit chi-square seems to have led to several uncritical applications of the same in actual practice. This fact has been noticed by many authors, e.g., Rao and Majumdar [4]. It turns out that in many such cases the mistake could have been avoided if a suitable correction were applied to Pearson's chi-square. For example, if the sample frequencies arise in random sampling from a finite population, without replacement, the only correction required is a multiplication by the reciprocal of the usual finite population correction to the variance of the sample mean, that is, by 1/(1-f), where f is the sampling fraction. Corrections necessary in another group of problems are obtained in the present paper. It is felt that the experimenter who is exposed to several alternative forms for the goodness-of-fit chi-square will be more cautions, as he has to think carefully before finally accepting any one of them (or possibly rejecting all of them). The development in Section 4 depends heavily on Rao [3], with which the reader is familiar.

2. CENTRAL LIMIT THEOREM FOR RANDOM SUMS

Let $X_1, X_2, \ldots$ be a sequence of independent identically distributed (i.i.d.) random variables with finite mean μ and finite variance σ^2. For each $\nu = 1,2,\ldots$, let n_ν denote a non-negative, integer-valued random variable which is independent of the X's and has finite mean a_ν and finite variance b_ν^2. Put

$$S_n = X_1 + X_2 + \ldots + X_n;$$

$$c_\nu = +(\mu^2 b_\nu^2 + a_\nu \sigma^2)^{1/2};$$

and $$Z_\nu = (S_{n_\nu} - \mu a_\nu)/c_\nu .$$

We shall now prove the following theorem:

Theorem 2.1

Under any of the following four conditions

$$\text{Prob}\left\{Z_\nu \le t\right\} \to \int_{-\infty}^{t} \frac{1}{\sqrt{2\pi}} e^{-u^2/2}\, du$$

as $\nu \to \infty$; this will be denoted symbolically by $\alpha(Z_\nu) \to N(0,1)$.

(1) $\mu = 0$, $\lim a_\nu = \infty$, $\text{plim}\; n_\nu/a_\nu = 1$

(2) $\lim a_\nu = \infty$, $\lim b_\nu^2/a_\nu = 0$

(3) $\lim a_\nu = \infty$, $\lim b_\nu^2/a_\nu = \lambda^2 < \infty$, $\alpha[(n_\nu - a_\nu)/b_\nu] \to N(0,1)$

(4) $\mu \neq 0$, $\lim b_\nu^2/a_\nu = \infty$, $\lim b_\nu^2/a_\nu^2 = 0$[(1)],

$\alpha[(n_\nu - a_\nu)/b_\nu] \to N(0,1)$

Proof: The characteristic functions of Z_ν and X_1 is easily seen to be given by

$$\Psi_\nu(\theta) = E\, e^{i\theta Z_\nu} = E\, f_\nu(n_\nu, \theta),$$

---

(1) For the purpose of proving theorem 2.1 it suffices to assume $\lim c_\nu = \infty$. We put in a stronger condition here as this will be required elsewhere.

where $f_\nu(n_\nu,\theta) = E\{e^{i\theta Z_\nu}|n_\nu\} = e^{-\frac{i\theta a_\nu \mu}{c_\nu}} [\Phi(\frac{\theta}{c_\nu})]^{n_\nu}$ (2.1)

and $\Phi(\theta) = E\, e^{i\theta X_1} = \exp[i\mu\theta - \frac{\theta^2}{2}\{\sigma^2+\varepsilon(\theta)\}]$, $\varepsilon(\theta) \to 0$ (1)

as $\theta \to 0$. Observe further that since $f_\nu(n_\nu,\theta)$ is itself a characteristic function, then

$$|f_\nu(n_\nu,\theta)| \leq 1 \tag{2.2}$$

Proof of (1):

Here $f_\nu(n_\nu,\theta) = \exp[-\frac{n_\nu\theta^2}{2a_\nu\sigma^2}\{\sigma^2 + \varepsilon(\frac{\theta}{\sigma\sqrt{a_\nu}})\}]$

and $\text{plim}\, f\,(n_\nu,\theta) = e^{-\theta^2/2}$. Hence using (2.2) and the dominated convergence theorem (Loeve [2], p. 125)

$$\lim \Psi_\nu\,(\theta) = e^{-\theta^2/2}\,.$$

Proof of (2):

Put

$$g_\nu(n_\nu,\theta) = \exp[i\theta\mu(\frac{n_\nu - a_\nu}{c_\nu}) - \frac{(n_\nu - a_\nu\{\sigma^2+\varepsilon(\theta/c_\nu)\}\theta^2}{2c_\nu^2}] \tag{2.3}$$

$$h_\nu(\theta) = \exp[-\frac{a_\nu\{\sigma^2+\varepsilon(\theta/c_\nu)\}\theta^2}{2c_\nu^2}]\,. \tag{2.4}$$

Observe that $\lim h_\nu(\theta) = e^{-\theta^2/2}$,

and $P\{|\frac{n_\nu - a_\nu}{c_\nu}| > \varepsilon\} \leq \frac{b_\nu^2}{\varepsilon^2 c^2} \to 0$ as $\nu \to \infty$.

Hence $\text{plim} \log_e g_\nu(n_\nu,\theta) = 0 => \text{plim}\, g_\nu(n_\nu,\theta) = 1$

$=> \text{plim}\, f_\nu(n_\nu,\theta) = e^{-\theta^2/2}$ since $f_\nu(n_\nu,\theta) = h_\nu(\theta)g_\nu(n_\nu,\theta)$; (2.2) and the dominated convergence theorem establishes that

$$\lim \Psi_\nu(\theta) = e^{-\theta^2/2}\,.$$

Proof of (3):

In this case,

$$\alpha[(n_\nu - a_\nu)/c_\nu] \to N(0, \frac{\lambda^2}{\mu^2\lambda^2 + \sigma^2}); \tag{2.5}$$

$$\text{plim}(n_\nu - a_\nu)/c_\nu^2 = 0, \text{ since } \lim c_\nu = \infty, \text{ while} \tag{2.6}$$

$$\lim a_\nu/c_\nu^2 = \frac{1}{\mu^2\lambda^2+\sigma^2} \text{ so that } h_\nu(\theta) \to \exp[-\frac{\theta^2\sigma^2}{2(\sigma^2+\mu^2\lambda^2)}]. \tag{2.7}$$

Since $h_\nu(\theta)$ is bounded for large ν and fixed θ by a constant, (2.6) and (2.7) = >

$$\text{plim}[f_\nu(n_\nu,\theta) - \tilde{f}_\nu(n_\nu,\theta)] = 0,$$

where $\tilde{f}_\nu(n_\nu,\theta) = \exp[i\theta\mu(\frac{n_\nu - a_\nu}{c_\nu}) - \frac{\theta^2\sigma^2}{2(\mu^2\lambda^2+\sigma^2)}]$.

Also $| \tilde{f}_\nu(n_\nu,\theta) | \leq 1$. Hence by the dominated convergence theorem

$$\lim E[(f_\nu(n_\nu,\theta) - \tilde{f}_\nu(n_\nu,\theta)] = 0$$

(2.5) establishes that $\lim \Psi_\nu(\theta) = \lim E[\tilde{f}_\nu(n_\nu,\theta)] = e^{-\theta^2/2}$.

Proof of (4):

In this case (2.6) holds, while in place of (2.5) and (2.7), we have

$$\alpha[(n_\nu - a_\nu)/c_\nu] \to N(0,\frac{1}{\mu^2}), \tag{2.8}$$

$$\text{and} \quad \lim a_\nu/c_\nu^2 = 0 . \tag{2.9}$$

Observing that $h_\nu(\nu) = 1$, and proceeding as in (3) we have

$$\lim \Psi_\nu(\theta) = e^{-\theta^2/2} .$$

Multivariate extension of theorem 2.1 is obtained on the usual lines, based on the following proposition due to Wald and Wolfowitz and Varadarajan, quoted and used by Rao [5], pp. 108-9.

"Let $\{\underset{\sim}{X}_n\}$ be a sequence of k-dimensional random variables. A necessary and sufficient condition that the cdf of $\underset{\sim}{X}_n$ converges to a cdf F is that the cdf of $\underset{\sim}{\lambda}' \underset{\sim}{X}_n$ converges to a limit for each fixed vector $\underset{\sim}{\lambda}$."

Thus we have

Theorem 2.2.

Let $\underset{\sim}{X}_1, \underset{\sim}{X}_2, \ldots$ be a sequence of independent identically distributed k-dimensional random variables with a finite mean vector $\underset{\sim}{\mu}$ and dispersion matrix Σ and let n_ν be defined as in theorem 2.1. Put

$$\underset{\sim}{S}_n = \underset{\sim}{X}_1 + \underset{\sim}{X}_2 + \ldots + \underset{\sim}{X}_n$$

$$\underset{\sim}{Z}_\nu = (S_{n_\nu} - a_\nu \mu)/\sqrt{a_\nu}$$

under conditions (1) - (4) of theorem 2.1, as $\nu \to \infty$, the following limiting distributions hold true:

Condition	r.v.	Limiting Distribution
1	Z_ν	$N_k(\underset{\sim}{0}, \Sigma)$
2	Z_ν	$N_k(\underset{\sim}{0}, \Sigma)$
3	Z_ν	$N_k(\underset{\sim}{0}, \lambda^2 \underset{\sim}{\mu}\, \underset{\sim}{\mu}' + \Sigma)$
4	$\begin{matrix} \underset{\sim}{p}' \\ P' \end{matrix} Z$ where $p = \dfrac{\sqrt{a_\nu}}{b_\nu \sqrt{\underset{\sim}{\mu}'\underset{\sim}{\mu}}} \underset{\sim}{\mu}$, P of rank k-1, such that $P' \underset{\sim}{\mu} = \underset{\sim}{0}$	$N_k(\underset{\sim}{0}, \Lambda)$ where $\Lambda = \begin{pmatrix} \underset{\sim}{\mu}' \underset{\sim}{\mu} & 0 \\ 0 & P'\Sigma P \end{pmatrix}$

Corollary 2.2

If Σ is p.d., as $\nu \to \infty$

$$(\underset{\sim}{S}_{n_\nu} - a_\nu \underset{\sim}{\mu})' (b_\nu^2 \underset{\sim}{\mu}\, \underset{\sim}{\mu}' + a_\nu \Sigma)^{-1} (\underset{\sim}{S}_{n_\nu} - a_\nu \underset{\sim}{\mu}) \tag{2.10}$$

is distributed as chi-square on k d.f., under any one of conditions (1) - (4).

Before proceeding further let us dwell a little on conditions (1) - (4), which appear to be more severe than what is required by contemporary authors to establish a central limit theorem for random sums (see Renyi [7] and [1], p. 258). Notice, however, that the location and scale corrections to the random sum, used by these authors, depend explicitly on the random sample size n_ν, which in our context is an unobserved random variable.

Examples of distributions of random sample size for which theorems 2.1 and 2.2 are true(2):

(i) $n_\nu \sim p(\lambda_\nu)$. Poisson variables with $\lambda_\nu \to \infty$

(ii) $n_\nu \sim b(N_\nu, p)$. Positive binomial variables with $N_\nu \to \infty$

(iii) $n_\nu \sim nb(k_\nu, \rho)$. Negative binomial variables with $k_\nu \to \infty$

(iv) $n_\nu \sim h(N_\nu, p, m_\nu)$. Positive hypergeometric variables with $m_\nu = N_\nu f$ and $N_\nu \to \infty$

(v) $n_\nu \sim nh(N_\nu, N_\nu p, m_\nu)$. Negative hypergeometric variables with(3) $m_\nu = N_\nu f$ and $N_\nu \to \infty$

(vi) truncated forms (truncation below) of (i) - (v)

3. TEST OF GOODNESS OF FIT OF A COMPOUND MULTINOMIAL

The vector of observed frequencies in a compound multinomial can be trivially expressed as a random sum of i.i.d. random vectors. Hence if E(n) is large and the distribution of n belongs to a family for which one of conditions (1) - (4) is true, expression (2.10), which in this case simplifies to

$$\sum_{1}^{k} \frac{(O_i - E_i)^2}{E_i} + c \, \frac{(O_. - E_.)^2}{E_{k+1}} , \tag{3.1}$$

has approximately a chi-square distribution on k d.f., where in (3.1),

$$E_i = \pi_i E(n), \; i = 1(1)k, \; E_{k+1} = (1-\pi_.) \, E(n) \tag{3.2}$$

$$\pi_. = \sum_{1}^{k} \pi_i , \quad O_. = \sum_{1}^{k} O_i , \quad E_. = \sum_{1}^{k} E_i ,$$

and $$c = \frac{[V(n) - E(n)] \, (1-\pi_.)}{[V(n) - E(n)] \, (1-\pi_.) - V(n)} .$$

---

(2) For an explanation of notations see Rao, Mitra and Matthai [6], p. 4.

(3) Frequency function

$$\binom{N_\nu p + x - 1}{x} \binom{N_\nu (1-p) + m_\nu - x - 1}{m_\nu - x} \Big/ \binom{N_\nu + m_\nu - 1}{m_\nu} ,$$

$x = 0(1)m_\nu$ (i.e., $x = 0, 1, \ldots, m_\nu$).

The factor c for some of the standard distributions is given below:

Distribution of n	Notation	Expression for c
Poisson	$p(\lambda)$	0
Positive binomial	$b(N,p)$	$\frac{(1-\pi_.)p}{1-p\pi_.}$
Negative binomial	$nb(k,\rho)$	$-\frac{(1-\pi_.)\rho}{1+\rho\pi_.}$
Hypergeometric	$h(N,Np,m)$	$\frac{[(N-1)-(N-m)(1-p)](1-\pi_.)}{(N-1)(1-\pi_.)+(N-m)(1-p)\pi_.}$
Negative hyper.	$nh(N,Np,m)$	$\frac{[N+1-(N+m)(1-p)](1-\pi_.)}{(N+1)(1-\pi_.)+(N+m)(1-p)\pi_.}$

Statistic (3.1) is useful in testing a hypothesis which completely specifies the compound multinomial distribution (that is, one which assigns numerical values to the class probabilities of the multinomial and in addition determines fully the distribution of n, the number of trials). It is interesting to record an alternative form of (3.1) as

$$\sum_1^k \frac{(O_i-O_.\pi_i')^2}{E_i} + \frac{(O_.-E_.)^2}{E(n)\pi_.(1-\pi_.)+V(n)\pi_.^2}, \qquad (3.3)$$

where $\pi_i' = \pi_i/\pi_.$, $i = 1(1)k$, for (3.3) is indeed an orthogonal[4] partition of chi-square into two components. The first has k-1 d.f. and the second 1 d.f. Since conditions (1) - (4) imply that for each i, plim $\{O_.\pi_i'/E_i\} = 1$, one might consider the following modified expression essentially equivalent to (3.3).

$$\sum_1^k \frac{(O_i-O_.\pi_i')^2}{O_.\pi_i'} + \frac{(O_.-E_.)^2}{E(n)\pi_.(1-\pi_.)+V(n)\pi_.^2}. \qquad (3.4)$$

[4] Note that $E(O_i-O_.\pi_i') = 0$. That a linear combination $\underset{\sim}{p}'\underset{\sim}{O}$ of observed frequencies with zero expectation is uncorrelated with $O_.$ is seen from the following argument. Let the probability of the ith class be $\alpha\pi_i$, $i = 1(1)k$, where α is an unknown parameter and π_i''s are known. Note that $O_.$ is complete sufficient for α and $E(\underset{\sim}{p}'\underset{\sim}{O}) \underset{\alpha}{\equiv} 0$.

This suggests that, while the first component with (k-1) d.f. tests the multinomial part of the hypothesis, the second one looks for deviations in E(n) from the value assigned to it by the hypothesis, though such deviations to a certain extent get mixed up with deviations in π .

4. ESTIMATION OF PARAMETERS

In this section we consider the problem of estimating parameters occurring in the specification of class probabilities. For simplicity we shall be concerned with the situation where the class probabilities $\pi_i(\theta)$ are known functions of a single unknown parameter θ; the multiparameter case is similar except for details. The functions $\pi_i(\theta)$ will be assumed to possess continuous first and second order derivatives with respect to θ and to satisfy $0 < c^2 < \pi_i(\theta)$, $\Sigma\ \pi_i(\theta) < 1$. Further, the distribution n_ν will be assumed to be known and to satisfy the conditions of theorem 2.1. Let $O_1, O_2, \ldots, O_k$ be the observed frequencies in k classes of a multinomial distribution with n_ν trials, and class probabilities $\pi_1(\theta), \pi_2(\theta), \ldots, \pi_k(\theta)$, respectively. $p_i = O_i/a$ provides a natural estimate of $\pi_i(\theta)$.

Definition: A statistic $t(p_1, p_2, \ldots, p_k)$, or briefly $t(\underset{\sim}{p})$, which is a function of the estimated relative frequencies $p_1, p_2, \ldots, p_k$ only, is said to be Fisher-consistent in an extended sense (EFC) for θ, if when the true proportions $\pi_i(\theta)$ are substituted for p_i, the function identically reduces to θ, i.e., if $t(\pi_1(\theta), \pi_2(\theta), \ldots, \pi_k(\theta)) = \theta$.

Since the conditions of theorem 2.1 imply plim $n_\nu = \infty$, plim $n_\nu/a_\nu = 1$, and therefore plim $p_i = \pi_i(\theta)$, it is clear that if $t(\underset{\sim}{p})$ is a continuous function of p_i and is EFC for θ, then plim $t(\underset{\sim}{p}) = \theta$. Lemma 4.1 can be proved on the same lines as in Lemma 2 of Rao [3].

Lemma 4.1.

If $t(\underset{\sim}{p})$ is EFC for θ and in addition admits continuous first partial derivatives with respect to θ_i, then

(i) $\alpha[\{t(\underset{\sim}{p}) - \theta\}/\sqrt{V_\nu}\ (t)] \to N(0,1)$.

(ii) The asymptotic variance $V_\nu(t)$ is given by

$$[a_\nu \sum_1^k \pi_i (\frac{\delta t}{\delta \pi_i})^2 + (b_\nu^2 - a_\nu) \{\sum_1^k \pi_i (\frac{\delta t}{\delta \pi_i})\}^2] / a_\nu^2$$

(iii) $V_\nu(t) \geq 1/[a_\nu \{\sum_1^k (\frac{\delta \log \pi_i}{\delta \theta})\}^2 \pi_i +$

$$c(\frac{\delta \log \pi_{k+1}}{\delta \theta})^2 \pi_{k+1}] = 1/a_\nu J_\nu(\theta),$$

where c is as defined in (3.2) and $\pi_{k+1}(\theta) = 1 - \pi_.(\theta)$.

4.1. The Method of Minimum Chi-Square

The minimum chi-square equation is obtained when expression (3.1) is differentiated with respect to θ and the derivative is equated to zero. We have thus

$$\eta(\underset{\sim}{p},\theta) = \sum_1^k \frac{p_i^2}{\pi_i^2} \frac{\delta \pi_i}{\delta \theta} - A(p_.,\pi_.) \frac{\delta \pi_.}{\delta \theta} = 0, \tag{4.1.1}$$

where $\underset{\sim}{p}$ represents the column vector $(p_1,p_2,\ldots,p_k)$,

$$A(p_.,\pi_.) = \frac{p_.^2}{\pi_.^2} - \frac{(p_. - \pi_.)^2}{(\pi_. + d_\nu \pi_.^2)} (1+2d_\nu \pi_.) \frac{2(p_. - \pi_.)}{\pi_. + d_\nu \pi_.^2},$$

$$p_. = \sum_1^k p_i, \quad d_\nu = (b_\nu^2 - a_\nu)/a_\nu.$$

Let θ_o represent the true value of parameter θ, $\pi_i{}^o = \pi_i(\theta_o)$ and $\underset{\sim}{\pi}^o$ represent the column vector $(\pi_1^o,\pi_2^o,\ldots,\pi_k^o)$.

Theorem 4.1.1.

For every $\delta > 0$, there exists an $\varepsilon > 0$ such that $|\underset{\sim}{p} - \underset{\sim}{\pi}^o| < \varepsilon \Rightarrow$

(i) There exists one and only one root $\hat{\theta}$ of minimum χ^2 equations (4.1.1) which differs from θ_o by less than δ. This root considered as a function of $\underset{\sim}{p}$ is continuous at $\underset{\sim}{\pi}^o, \hat{\theta}(\underset{\sim}{\pi}^o) = \theta_o$.

(ii) $\hat{\theta}$ is Frechet differentiable. $\hat{\theta} - \theta_o$ can be approximated by the linear functional $L(\underset{\sim}{p},\underset{\sim}{\pi}^o) =$

$$\sum \frac{p_i - \pi_i^o}{\pi_i^o} (\frac{\delta \pi_i}{\delta \theta})_o - \frac{(p_. - \pi_.^o)\, d_\nu \pi_.^o}{\pi_.^o + d_\nu \pi_.^2} - (\frac{\delta \pi_.}{\delta \theta})_o.$$

(iii) $\alpha[\sqrt{a_\nu J_\nu(\theta_o)}\, (\hat{\theta} - \theta_o)] \to N(0,1)$, as $\nu \to \infty$.

$$\text{(iv)}\quad \sum_{1}^{k} \frac{(O_i - \hat{E}_i)^2}{\hat{E}_i} + c\,\frac{(O_{.} - \hat{E}_{.})^2}{\hat{E}_{k+1}}$$

is, in the limit, distributed as chi-square on k-1 d.f., where

$$\hat{E}_i = a_\nu\, \pi_i(\hat{\theta}),\ E_{.} = \Sigma\, E_i,\ E_{k+1} = a_\nu - \hat{E}_{.}\ .$$

Note: The only exception to the theorem is when condition (4) holds and further $\frac{\delta \log \pi_i}{\delta\theta}$ is a constant possibly depending on θ but not on i.

Proof of (i):

The proof consists in noting that

$$\frac{\delta \eta(\underset{\sim}{p},\theta)}{\delta\theta} + 2J_\nu(\theta_o) \to 0 \tag{4.1.2}$$

as $\underset{\sim}{p} \to \underset{\sim}{\pi}^o$ and $\theta \to \theta_o$, where $J_\nu(\theta)$ is as defined in Lemma 4.1 (iii) and that

$$J(\theta_o) = \sum_{1}^{k} \left\{ \left(\frac{\delta \log \pi_i}{\delta\theta}\right)_o - \left(\frac{\delta \log \pi_{.}}{\delta\theta}\right)_o \right\}^2 \pi_i^o +$$

$$\frac{\pi_{k+1}^o}{\pi^o(1-\pi^o) + \frac{b_\nu^2}{a_\nu}\, \pi_{.}^{o\,2}} \left(\frac{\delta \log \pi_{k+1}}{\delta\theta}\right)^2 \pi_{k+1}^o$$

is positive and remains bounded away from zero in cases excluded by the exception noted above. The rest of the proof follows as in the proof of corresponding results of Rao [3].

Proof of (ii):

Equation (4.1.1) with $\hat{\theta}$ the selected root is

$$0 = \eta(\underset{\sim}{p},\hat{\theta}) = \eta(\underset{\sim}{p},\theta_o) + (\hat{\theta} - \theta_o)\, \eta'(\underset{\sim}{p},\tilde{\theta})$$

where $\tilde{\theta} \to \theta_o$ as $\underset{\sim}{p} \to \underset{\sim}{\pi}^o$. Thus

$$\hat{\theta} - \theta_o = \eta(\underset{\sim}{p},\theta_o)/\eta'(\underset{\sim}{p},\tilde{\theta}) \tag{4.1.3}$$

Consider $d(\underset{\sim}{p}, \underset{\sim}{\pi}^o) = \max \left\{ |p_1 - \pi_1^o|, \ldots, |p_k - \pi_k^o|, |p_{.} - \pi_{.}^o| \right\}$

and

$$z = \frac{1}{d(\underset{\sim}{p},\underset{\sim}{\pi}^o)} \{(\hat{\theta} - \theta_o) - L(\underset{\sim}{p},\underset{\sim}{\pi}^o)\}. \tag{4.1.4}$$

Substituting (4.1.3) in (4.1.4) and noting that $\eta(\underset{\sim}{\pi}^{o},\underset{\sim}{\theta}_{o}) = 0$, we have

$$z = \frac{1}{d(\underset{\sim}{p},\underset{\sim}{\pi}^{o})} \left\{ \frac{\eta(\underset{\sim}{p},\theta_{o}) - \eta(\underset{\sim}{\pi}^{o},\theta_{o})}{\eta'(\underset{\sim}{p},\tilde{\theta})} - L(\underset{\sim}{p},\underset{\sim}{\pi}^{o}) \right\}.$$

Hence

$$|z| \leq \Bigg| \frac{1}{\eta'(\underset{\sim}{p},\tilde{\theta})} \Sigma \frac{p_i + \pi_i^{o}}{\pi_i^{o\,2}} \left(\frac{\delta\pi_i}{\delta\theta}\right)_{o} -$$

$$\left\{ \frac{2(p_{.} + {}^{o}_{.})}{\pi_{.}^{o\,2}} + \frac{2}{\pi_{.}^{o} + d_{\nu}\pi_{.}^{o\,2}} \right\} \left(\frac{\delta\pi_{.}}{\delta\theta}\right)_{o} - L(\underset{\sim}{p},\underset{\sim}{\pi}^{o}) \Bigg|,$$

which in view of (4.1.2) $\to 0$ as $d(\underset{\sim}{p},\underset{\sim}{\pi}^{o}) \to 0$. This completes proof of (ii).

Proof of (iii):

(iii) follows from (ii) once it is noted that

$$\text{plim} \sqrt{a_{\nu} J_{\nu}(\theta_{o})}\; z = 0,$$

and one appeals to Slutsky's Theorem.

Proof of (iv):

To prove (iv) we write

$$\sum_{1}^{k} \frac{(O_i - E_i^{o})^2}{E_i^{o}} + c \frac{(O_{.} - E_{.}^{o})^2}{E_{k+1}^{o}} = \sum_{1}^{k} \frac{(O_i - \hat{E}_i)^2}{\hat{E}_i} + c \frac{(O_{.} - \hat{E}_{.})^2}{\hat{E}_{k+1}} -$$

$$(\theta_{o} - \theta) a_{\nu} \eta(p,\hat{\theta}) - \frac{(\theta_{o} - \hat{\theta})^2}{2} \left(\frac{\delta\eta(\underset{\sim}{p},\theta)}{\delta\theta}\right)_{\tilde{\theta}},$$

where $E_i^{o} = a_{\nu}\pi_i(\theta_{o})$, $E_{.}^{o} = \Sigma E_i^{o}$, $E_{k+1}^{o} = a_{\nu} - E_{.}^{o}$, and $\tilde{\theta} \to \theta_{o}$ as $\underset{\sim}{p} \to \underset{\sim}{\pi}^{o}$. Result (iv) follows from results (ii) and (iii), since $\eta(\underset{\sim}{p},\hat{\theta}) = 0$ and $\left(\frac{\delta\eta(\underset{\sim}{p},\theta)}{\delta\theta}\right)_{\tilde{\theta}} + 2J_{\nu}(\theta_{o}) \to 0$ as $\underset{\sim}{p} \to \underset{\sim}{\pi}^{o}$.

5. NUMERICAL ILLUSTRATION

The female knapweed gall-fly Urophara jaceana lays its eggs in batches within unopened flower heads of black knapweed

Centaurea nemoralis. The second instar larva hatches from the egg, moves a short distance within the flower head, and produces a hard gall-cell in which occur all further stages of development up to emergence of the adult fly. An examination of a one-square-meter sample of knapweed over a period led to the following figures.(5)

Number (n) of eggs laid	:	Not available
number of infertile eggs	:	not available
number of larvae dying before forming galls	:	37
number of larvae killed by parasites	:	66
number of larvae developing into adult flies	:	79

Examine if the data is consistent with the hypothesis which specifies the proportion in the four classes as 10%, 20%, 35% and 35%, respectively. Mean value of n can be taken to be 200 and standard deviation 20.

$$\pi_{.} = 0.20 + 0.35 + 0.35 = 0.90$$

$$c = \frac{0.1(400-200)}{0.1(400-200) - 400} = -\frac{20}{380}$$

i	O_i	E_i	difference	$(O_i-E_i)^2/E_i$
1	37	40	-3	.225
2	66	70	-4	.229
3	79	70	9	1.157
Total	182	180	2	1.611
		$c(O_{.}-E_{.})^2/E_{k+1}$	=	- 0.001
		Value of χ^2	=	1.610

The upper 5% value of χ^2 on 3 d.f. being 7.81 the observed value of χ^2 is not significant. Hence the data do not reveal significant deviations from the hypothesis.

(5) data adapted from Varley [9]

ACKNOWLEDGMENT

The author wishes to thank Dr. B. Ramachandran for his help in Section 2 of this paper.

REFERENCES

[1] Feller, W. 1966. An Introduction to Probability Theory and Its Application, Vol. II. John Wiley and Sons, New York.

[2] Loeve, M. 1960. Probability theory, second edition. D. Van Nostrand Company, Princeton, N.J.

[3] Rao, C. R. 1955. Theory of the method of estimation by minimum chi-square, Bull. Int. Stat. Inst. 35:25-32.

[4] ————, and Majumdar, D. N. 1958. Bengal Anthropometric Survey, 1945 - A Statistical Study, Sankhyā 19:201-408.

[5] ————. 1965. Linear Statistical Inference and Its Applications. John Wiley and Sons, New York.

[6] ————; Mitra, S. K.; and Matthai, A. 1966. Formulae and Tables for Statistical Work. Statistical Publishing Society, Calcutta.

[7] Renyi, A. 1960. On the central limit theorem for the sum of a random number of independent random variables, Acta Math. 11:97-102.

[8] Sibuya, M.; Yoshimura, I.; and Shimizu, R. 1964. Negative multinomial distribution, Ann. Inst. Stat. Math. 16:409-26.

[9] Steyn, H. S. 1955. On discrete multivariate probability functions of hypergeometric type, Nederl. Akad. Wetensch. Proc., Series A, 58:588-95.

[10] ————. 1959. On χ^2 tests for contingency tables of negative binomial type, Statistica Neerlandica, 13:433-44.

[11] Varley, G. C. 1947. The natural control of population balance in the knapweed gall-fly (Urophora jaceana), Journal of Animal Ecology 16:139-87.

Note added in proof: The following papers which discuss the multinomial distribution when the parameter n has a power series distribution have been brought to the attention of the author:

[1] Patil, G. P. 1968. On sampling with replacement from populations with multiple characters, Sankhyā Series B 30:355-66.

[2] Joshi, S. W.; and Patil, G. P. 1968. Certain structural properties of the sum-symmetric power series distributions, Technical Report and Preprint No. 5. The Pennsylvania State University, Department of Statistics.

[3] ———. 1968. Sum-symmetric power series distributions and minimum variance unbiased estimation, Technical Report and Preprint No. 5. The Pennsylvania State University, Department of Statistics.

NORMAL APPROXIMATIONS TO THE POISSON DISTRIBUTION*

W. MOLENAAR
Mathematisch Centrum
Amsterdam

SUMMARY

This paper considers improvements, some known and some possibly new, on the classical approximation to the Poisson distribution. The emphasis is on approximations which are simple enough for hand calculation and are accurate near the customary significance levels. The evaluation of accuracy and simplicity is to some extent a matter of taste.

Let F_λ denote the (cumulative) Poisson distribution function with parameter λ (see Section 1). In the equation $F_\lambda(k) = P$ one can try to

(1) find P from k and λ (distribution problem, Section 1),
(2) find λ from k and P (confidence problem, Section 2),
(3) find k from λ and P (fractile problem, Section 3).

In all three cases, the transcendental equation admits no explicit solution, but asymptotic expansions for the solution exist. They are used in the Appendix for a comparison of approximate solutions based on the normal distribution. The paper briefly reports on numerical investigations of accuracy.

A general advice for the three problems mentioned above is given in Tables 1, 2, and 3, respectively. The choice of an approximation will depend on the desired accuracy and the available computational facilities.

---

*Report S-412, Statistische Afdeling, Mathematisch Centrum, Amsterdam.

1. THE DISTRIBUTION PROBLEM

Throughout this paper λ is a positive real number, k a nonnegative integer, and P a real number with $0 < P < 1$. Let F_λ denote the Poisson distribution function:

$$F_\lambda(k) = \sum_{j=0}^{k} e^{-\lambda} \lambda^j/j! \ , \qquad (1)$$

and let Φ denote the unit normal distribution function:

$$\Phi(u) = (2\pi)^{-\frac{1}{2}} \int_{-\infty}^{u} \exp(-\tfrac{1}{2}t^2)\ dt \ . \qquad (2)$$

As tables of F are not always immediately accessible and interpolation in these tables is cumbersome, our goal is to find simple functions u of k and λ such that $F_\lambda(k) \simeq \Phi(u)$. We shall use u as a general notation for such a deviate, and v, w, w_1, etc., for special cases. The simplest example is the standardized variable $w_1 = (k-\lambda)\lambda^{-\frac{1}{2}}$, to which one usually adds a continuity correction:

$$w = (k + \tfrac{1}{2} - \lambda)\lambda^{-\frac{1}{2}} \ . \qquad (3)$$

Another example is the square root transform, used for variance stabilization; with a suitable continuity correction it becomes

$$v = 2(k+1)^{\frac{1}{2}} - 2\lambda^{\frac{1}{2}} \ . \qquad (4)$$

Our advice for normal approximations to $F_\lambda(k)$ is summarized in Table 1. The problem of finding accurate "normal deviates" was studied in a more general context by Peizer and Pratt [7]. Their proposal, listed as v^{***} in Table 1, is somewhat laborious for hand calculation. Our evaluation of simplicity is based on a preference for approximations requiring no more than square roots and a cumulative normal table. The use of third roots, natural logarithms and other special functions is unattractive in "quick and dirty" work. With an electronic computer, direct summation of (1) for, say, $\lambda < 30$, and calculation of $\Phi(v^{***})$ for larger λ, guarantees accuracy to 5 or 6 decimal places when a good approximation for Φ is used.

The choice of approximating deviates in Table 1 is based on theoretical considerations summarized in the Appendix and on a numerical investigation of errors, which includes values of λ ranging from 0.5 to 300, and, for each λ, all values of k satisfying $0.001 \leq F_\lambda(k) \leq 0.999$. For us, accurate approximation near the customary significance levels is the most important. It is hardly ever interesting to be quite sure that a probability is 3.10^{-4} and not 4.10^{-4}, so small tails were not considered.

A brief report on the numerical errors is not an easy task. The errors $\Phi(u) - F_\lambda(k)$ will be sketched as a function of k for fixed λ. Figures 1, 2, 3, and 4 pertain to λ = 2, 10, 30 and 100, respectively, different curves belonging to different deviates, u. The horizontal scale gives k and, in italics, $F_\lambda(k)$; the vertical scale gives the error $\Phi(u) - F_\lambda(k)$. Though it is hardly possible to give information on, say, u and v^{***} in the same picture, the simple linear error scale is used in order to facilitate the interpretation of the pictures by the reader.

For example for λ = 10 (Fig. 2) and k = 5, the italics give $F_{10}(5) = 0.067$. The graph for v gives at k = 5 an error of +0.010. One thus finds that $\Phi(v) = 0.067 + 0.010 = 0.077$, an overestimation of the tail with a relative error of some 15 percent.

It is also interesting to give the <u>minimal</u> parameter value for which the <u>relative</u> error at the 0.05 and 0.95 points of the distribution is less than 1 percent, i.e., for which the approximation lies between 0.0495 and 0.0505 (between 0.9495 and 0.9505). This minimal λ exceeds 300 for v and w, lies between 7 and 10 for v^*, between 3 and 4 for v^{**}, and below 0.5 for v^{***}. For the 0.025 and 0.975 point the approximation lies between 0.02475 and 0.02525 (between 0.97475 and 0.97525) if λ exceeds a value that is more than 300 for w, and is about 70 for v (for v^*, v^{**}, v^{***}, as above). Whereas v is especially accurate near the 0.025 and 0.975 points, most other approximations have a relative error that is large for small tails and small for large tails. More details will be given in a subsequent publication.

2. THE CONFIDENCE PROBLEM

When k events have been observed, in a process where the number of events has a Poisson distribution with unknown expectation λ, the upper bound λ_2 with confidence level $1-\alpha$ is the solution of $F_{\lambda_2}(k) = \alpha$. The similar lower bound λ_1 satisfies

$1-F_{\lambda_1}(k-1) = \alpha$. Note that the confidence level $1-\alpha$ holds for each bound separately. Thus the random interval (λ_1,λ_2) contains the fixed unknown value λ with probability at least $1-2\alpha$; and when a confidence interval with confidence level $1-\alpha$ is desired, one should replace α by $\frac{1}{2}\alpha$ in all subsequent formulae. The exact value of the probability of incorrect statements depends on the value of λ, because of the discrete nature of the Poisson distribution.

Tables for λ_1 and λ_2 exist for small values of k and customary values of α. Moreover λ_2 equals the $1-\alpha$ fractile of a χ^2 distribution with 2k f.f. Nevertheless, when such tables are not available or do not contain the relevant values of k and/or α, approximations to confidence bounds become useful.

An advice for such approximations is given in Table 2. As before, there is a subjective element in the evaluation of accuracy and computational labor. Some mathematical background of the Table is given in the appendix. In a numerical investigation many expressions (some are listed in the appendix) were tried for $\alpha = 0.1$, 0.05, 0.025, 0.01 and 0.005. No choice was uniformly superior for all α and k.

In our short report on errors we compare the approximated bounds to the exact bounds, as a comparison of probabilities of correct statements becomes rather complicated. We shall not consider $k = 0$ for upper bounds nor $k = 0$ or $k = 1$ for lower bounds, since one easily finds the exact values to be $-\ln \alpha$, 0, and $-\ln(1-\alpha)$, respectively, in these cases.

The "quick work" suggestion of Table 2 differed less than 0.25 from the exact bound, with a few exceptions having error 0.34 or less for upper bounds with $\alpha = 0.1$.

The combined "more accurate" advice has an absolute difference between approximated and exact upper bounds of, at most, 0.05, provided that: $k \geq 1$ when $\alpha = 0.1$; $k \geq 5$ when $\alpha = 0.05$; $k \geq 12$ when $\alpha = 0.025$; $k \geq 8$ when $\alpha = 0.01$; $k \geq 4$ when $\alpha = 0.005$. For lower bounds the same statement holds for: all k when $\alpha = 0.1$ or $\alpha = 0.05$, $k \geq 7$ when $\alpha = 0.025$, $k \geq 5$ when $\alpha = 0.005$. As a rule this approximation is conservative; i.e., the approximated upper [lower] bounds are higher [lower] than the exact values. The only exceptions observed to this conservatism were lower bounds for $\alpha = 0.005$ $(k < 70)$ and $\alpha = 0.01$ $(k < 4)$.

Anderson and Burstein ([1], [2]) use $k - \xi k^{\frac{1}{2}} + R$ and the corresponding upper bound for all α, but we found that $\{(k+B)\}^{\frac{1}{2}} - \frac{1}{2}\xi\}^2$ was much better for $\alpha = 0.05$ and 0.1, and slightly better for $\alpha = 0.025$.

The "still better" formula of Table 2 was indeed nearly always better, the only exceptions being for lower bounds $3 \leq k \leq 11$ when $\alpha = 0.01$ and $k \geq 7$ when $\alpha = 0.005$, for upper bounds $k \geq 10$ when $\alpha = 0.005$. It has an error which rarely exceeded 0.02 and never exceeded 0.041 in our investigation.

The first "very accurate" suggestion has an error not exceeding 0.051, and for $\alpha \geq 0.025$ not exceeding 0.021. The second one differs never more than 0.007 from the exact confidence bound: it is better than the first one for small k and small α, and this is the region where the largest errors occur.

In the preceding statements about errors, k = 0 for upper bounds and k = 0 or 1 for lower bounds were not considered; the simple exact solution for these cases has already been mentioned.

3. THE FRACTILE PROBLEM

When a hypothesis is tested about an unknown Poisson parameter, it is frequently desired to determine k, from given α and λ, such that $F_\lambda(k) \leq \alpha < F_\lambda(k+1)$, or $F_\lambda(k-1) < 1-\alpha \leq F_\lambda(k)$. An equality sign, i.e., $F_\lambda(k) = \alpha$ or $= 1-\alpha$, is only possible for the very special values of λ which are an upper or lower $(1-\alpha)$ level confidence bound corresponding to some integer k.

Table 3 gives advice on normal approximations to Poisson P fractiles for given values of P and λ. The preceding remark indicates how one uses these fractiles when determining a rejection region. In many cases the probability of exceedance for an observed value (Section 1) will be more informative. Furthermore, the exact [approximated] fractile has a jump of +1 at the exact [approximated] corresponding confidence bound values for λ, so Section 2 gives some information on the errors. Therefore we shall be brief about fractile approximation and skip the technical complications arising from the discrete character of the distribution.

In our numerical investigation of approximations to the Poisson P fractile, we took P = 0.005, 0.01, 0.025, 0.05, 0.1, 0.9, 0.95, 0.975, 0.99, and 0.995; and λ had the special (confidence) values mentioned above. For these values, accuracy was measured by the absolute difference between approximated and exact k.

For the "quick work" suggestion of Table 3, this difference never exceeded 0.85 and rarely exceeded 0.4 for the ten values of P mentioned above. It was found that $\lambda + \xi\lambda^{\frac{1}{2}}$ was better than our "quick work" idea for P = 0.025 and all λ, P = 0.05 and $\lambda < 5$, P = 0.99 and $\lambda < 12$, P = 0.995 and $\lambda < 5$. As the "quick work" idea is defeated by $\lambda + \xi\lambda^{\frac{1}{2}} - \frac{1}{2}$ for roughly $.07 \leq P \leq .93$, it is only retained in Table 1 because it is simple and reasonably accurate for all P simultaneously.

The "more accurate" suggestion was found to give an error never exceeding 0.11 provided that $\lambda > 1$, never exceeding 0.03 provided that $\lambda > 33$. Unless P = 0.995 it was found to be conservative (to underestimate quantiles for P near 0, to overestimate them for P near 1).

The error of the "very accurate" formula was never found to be more than 0.02 provided that $P \geq 0.025$ and $\lambda > 1$, but for P = 0.01 or P = 0.005 it can be 0.08 even if $\lambda > 1$.

APPENDIX

Monotonicity considerations guarantee that for given λ and k there exists a unique "exact normal deviate" $\xi = \xi(k,\lambda)$ satisfying $F_\lambda(k) = \Phi(\xi)$. Its explicit solution from this equation is impossible, but for $\lambda \to \infty$ and bounded w or v (defined in equations (3) and (4)) ξ has the asymptotic expansions

$$\xi = w + \frac{1-w^2}{6\sqrt{\lambda}} + \frac{5w^3-2w}{72\lambda} + \frac{128+79w^2-249w^4}{6480\lambda\sqrt{\lambda}} + O(\lambda^{-2}) \; ; \qquad \text{(A.1)}$$

$$\xi = v + \frac{v^2-4}{12\sqrt{\lambda}} + \frac{10v-v^3}{72\lambda} + \frac{21v^4-371v^2-52}{6480\lambda\sqrt{\lambda}} + O(\lambda^{-2}) \; . \qquad \text{(A.2)}$$

The well-known expansion (A.1) can be derived by combining

$$F_\lambda(k) = \sum_{j=0}^{k} e^{-\lambda}\lambda^j/j! = \int_\lambda^\infty t^k e^{-t}/k! \, dt \qquad \text{(A.3)}$$

with the Cornish-Fisher [4] expansion valid for the gamma distribution. An alternative proof uses Cornish-Fisher directly for the Poisson distribution, with correction terms for its lattice character (Esseen [5], p. 61). The expansion (A.2) follows from (A.1) if w is written as a function of v and λ.

When (A.3) is combined with asymptotic normality of the gamma distribution, one finds the deviate $y = (k+1-\lambda)(k+1)^{-\frac{1}{2}}$. Riordan [8] gives the expansion ($k \to \infty$, y bounded)

$$\xi = y + \frac{y^2-1}{3\sqrt{k+1}} + \frac{7y^3-y}{36(k+1)} + \frac{219y^4-14y^2-13}{1620(k+1)\sqrt{k+1}} + O(k+1)^{-2}) \;, \qquad \text{(A.4)}$$

of which each partial sum is usually less accurate than the same number of terms from (A.1) or (A.2). If one uses (A.3) and the Wilson-Hilferty approximation to χ_n^2, which makes $\{(\chi^2/n)^{1/3} - 1 + \frac{2}{9n}\}\{\frac{2}{9n}\}^{-\frac{1}{2}}$ unit normal, one obtains a deviate

$$z = 3(k+1)^{\frac{1}{2}} - \frac{1}{3}(k+1)^{-\frac{1}{2}} - 3\{\lambda(k+1)^{\frac{1}{2}}\}^{1/3} \;, \qquad \text{(A.5)}$$

which is somewhat laborious (third root!), slightly superior to v^* for large λ, but hardly ever superior to v^{**}. The Fisher approximation to χ_n^2, which makes $(2\chi^2)^{\frac{1}{2}} - (2n-1)^{\frac{1}{2}}$ unit normal, gives the deviate $v_1 = 2(k+\frac{3}{4})^{\frac{1}{2}} - 2\lambda^{\frac{1}{2}}$, mentioned in Table 1 for quick work and probabilities between 0.06 and 0.94.

In Table 1 it was advised to use v and v_1, and never w, w_1, or y. For a motivation of this advice, observe that straightforward series expansions lead to

$$\begin{aligned}
y &= (k+1-\lambda)(k+1)^{-\frac{1}{2}} &&= \xi + (4-4\xi^2)/(12\sqrt{\lambda}) + O(\lambda^{-1}),\\
w &= (k+\tfrac{1}{2}-\lambda)\lambda^{-\frac{1}{2}} &&= \xi + (2\xi^2-2)/(12\sqrt{\lambda}) + O(\lambda^{-1}),\\
v_1 &= 2(k+\tfrac{3}{4})^{\frac{1}{2}}-2\lambda^{\frac{1}{2}} &&= \xi + (1-\xi^2)/(12\sqrt{\lambda}) + O(\lambda^{-1}), \qquad \text{(A.6)}\\
w_1 &= (k-\lambda)\lambda^{-\frac{1}{2}} &&= \xi + (2\xi^2-8)/(12\sqrt{\lambda}) + O(\lambda^{-1}),\\
v &= 2(k+1)^{\frac{1}{2}}-2\lambda^{\frac{1}{2}} &&= \xi + (4-\xi^2)/(12\sqrt{\lambda}) + O(\lambda^{-1}).
\end{aligned}$$

For large λ it follows that y, w, and v_1 are accurate near $\xi = \pm 1$ (tails of 0.16) and w_1 and v near $\xi = \pm 2$ (tails of 0.023). Far from these special values the error of v is $(-\frac{1}{2})$ times the error of w_1, etc. As Taylor expansion gives, e.g., $\Phi(v) - \Phi(\xi) = (v-\xi)\,\Phi'(\xi) + O(v-\xi)^2$, the error $\Phi(v) - F_\lambda(k)$ behaves like the error in the deviate and is about $(-\frac{1}{2})$ times

$\Phi(w_1) - F_\lambda(k)$. When all numerators which have denominator $12\sqrt{\lambda}$ in (A.6) are similarly examined as functions of ξ, our suggestions in Table 1 become apparent. The picture does not essentially change when also the $O(\lambda^{-1})$ terms in (A.6) are considered, and numerical results confirm our findings with only a few exceptions. A more complete discussion will be published elsewhere.

Next we consider approximations with error $O(\lambda^{-1})$. Similar asymptotic considerations establish the superiority of v^* over the first two terms of (A.1), (A.2) or (A.3), and over the Wilson-Hilferty result (A.5). The well-known $\Phi(w) + (1-w^2)$ $\Phi'(w)/(6\sqrt{\lambda})$ is usually more accurate than $\Phi(w + (1-w^2)/(6\sqrt{\lambda}))$, but also more laborious, and it is worse than $\Phi(v^*)$.

Finally v^{***}, v^{**} and the first three terms of the expansions for ξ all have error $O(\lambda^{-3/2})$, the advantage of v^{***} over v^{**} and of v^{**} over the others being a smaller coefficient of $\lambda^{-3/2}$ for most values of ξ.

The deviates v^* and v^{**} were found by expanding the general formula $2(k+1+b+ck^{-\frac{1}{2}})^{\frac{1}{2}} - 2(\lambda+\beta+\gamma\lambda^{-\frac{1}{2}})^{\frac{1}{2}}$, and choosing two or three of the expressions b, c, β, γ as zero, the other(s) as such functions of w that give agreement up to the highest possible order. The use of polynomials in v or v_1 instead of w turned out to be generally less accurate. Many other approximating deviates were tried, some new and some previously published. Though some of them will occasionally be good for special values of λ and k, the advice of Table 1 will be effective in most cases.

For an explanation of Section 2, observe that the $1-\alpha$ confidence bounds λ_2 and λ_1 both satisfy the expansion (Campbell, [3], for $h \to \infty$ and ξ fixed,

$$\lambda_i = h + \xi\sqrt{h} + \frac{\xi^2-1}{3} + \frac{\xi^3-7\xi}{36\sqrt{h}} + \frac{16-7\xi^2-3\xi^4}{810h} + O(h^{-3/2}) , \qquad \text{(A.7)}$$

if one takes $h = k+1$ and ξ such that $\Phi(\xi) = 1-\alpha$ for the upper bound, $h = k$ and ξ such that $\Phi(\xi) = \alpha$ for the lower bound. The expansion can be obtained from the inverse relation to (A.4) expressing y in terms of ξ and h. It also follows from the expansion for χ^2 fractiles (Goldberg and Levine, [6]). The first few terms suffice if h is large, convergence being slower for small α.

As both bounds λ satisfy $F_\lambda(h-1) = \Phi(-\xi)$ (cf. Section 2), one may put $-\xi \approx v = 2h^{\frac{1}{2}} - 2\lambda^{\frac{1}{2}}$, which gives $\lambda \approx (h^{\frac{1}{2}} + \frac{1}{2}\xi)^2$. With a table of squares and square roots this may be even simpler to compute than $h+\xi h^{\frac{1}{2}}$, and it is theoretically more accurate when its absolute error $|4-\xi^2|/12$ is less than $|\xi-1|/3$, i.e., when $|\xi| > \sqrt{1.6}$. In practice this means $\alpha \leq .1$ for lower bounds but $\alpha < .1$ for upper bounds, as there the $O(h^{-\frac{1}{2}})$ term of the error more or less cancels the $O(1)$ term for $h+\xi h^{\frac{1}{2}}$, but has the same $(h^{\frac{1}{2}}+\frac{1}{2}\xi)^2$.

Similar asymptotic considerations show the superiority of the first "more accurate" formula of Table 2 over $k - \xi k^{\frac{1}{2}} + R$ whenever $2 \leq |\xi| \leq 4$. The "still better" formula, following directly from the Wilson-Hilferty approximation to χ^2, has a smaller error (though also of order $h^{-\frac{1}{2}}$) unless $\sqrt{6}\ |\xi| < 3$. We observe that

$$h + \frac{\xi^2-1}{3} + \xi(h + \frac{\xi^2-7}{18})^{\frac{1}{2}} = \{(h + \frac{\xi^2-7}{18})^{\frac{1}{2}} + \frac{1}{2}\xi\}^2 + \frac{\xi^2+2}{36} \qquad \text{(A.8)}$$

is usually slightly superior to Wilson-Hilferty for large k. Inversion of the very accurate Peizer-Pratt deviate v^{***} (Table 1) is too complicated.

The first "very accurate" formula, consisting of the first five terms of (A.7), is good for large α (small ξ) because the first neglected term has a contribution proportional to $\xi^5 h^{-3/2}$. The second "very accurate" suggestion was found empirically by plotting the exact E satisfying $\lambda_1 = \{(k+B+CE^{-\frac{1}{2}})^{\frac{1}{2}} - \frac{1}{2}\xi\}^2$ as a function of k, and similarly for upper bounds. The value of E did not vary seriously with α. Though it is asymptotically equal to $k + O(k^{\frac{1}{2}})$, the formulae $\frac{7}{6}k+3$ and $\frac{3}{4}(k-1)$ give a better performance for small k, while for large k the whole term $CE^{-\frac{1}{2}}$ is so small that the wrong coefficient of k does no harm.

The Poisson P fractile treated in Section 3 satisfies the expansion (Campbell, [3]), for $\lambda \to \infty$ and ξ fixed

$$k = \lambda + \xi\lambda^{\frac{1}{2}} + \frac{\xi^2-4}{6} - \frac{\xi^3+2\xi}{72\lambda^{\frac{1}{2}}} + \frac{3\xi^4+7\xi^2-16}{810\lambda} + O(\lambda^{-3/2}), \qquad \text{(A.9)}$$

where ξ is the unit normal P fractile, i.e., $\Phi(\xi) = P$. This expansion follows from (A.1) when $w = (k+\frac{1}{2}-\lambda)\lambda^{-\frac{1}{2}}$ is expressed in terms of ξ and λ. It is obvious that the "quick work" suggestion of Table 3 is the equivalent of v in Table 1. It has an error with leading term $(\xi^2-4)/12$, against $(4-\xi^2)/6$ for the more customary $\lambda+\xi\lambda^{\frac{1}{2}}$. As before the second "more accurate" suggestion has a smaller asymptotic error than the first one for $2 < |\xi| < 4$. The Wilson-Hilferty and Peizer-Pratt deviates lead to a very complicated expression for the quantile.

Table 1

Advice for the choice of a normal deviate u such that $\Phi(u)$ is an approximation to the probability $F_\lambda(k)$ of k or less events in a Poisson distribution with expectation λ. For Φ and F_λ cf. Section 1, (1) and (2).

For quick work, use:

$v = 2(k+1)^{\frac{1}{2}} - 2\lambda^{\frac{1}{2}}$, for reasonable accuracy near the customary significance levels (for probabilities between 0.06 and 0.94 it is better to use $v_1 = 2(k+\frac{3}{4})^{\frac{1}{2}} - 2\lambda^{\frac{1}{2}}$);

For more accurate work, use:

$v^{*} = 2\{k + (w^2+8)/12\}^{\frac{1}{2}}$, where $w = (k+\frac{1}{2} - \lambda)\lambda^{-\frac{1}{2}}$

or, still better,

$$v^{**} = 2\{k + (w^2+5)/9\}^{\frac{1}{2}} - 2\{\lambda + (w^2-4)/36\}^{\frac{1}{2}} .$$

For results accurate at the P and 1 - P fractile, but rough elsewhere, one might use a modified form of v^{**}, viz.

$$v_P = 2\{k + (\xi^2+11)/18\}^{\frac{1}{2}} - 2\{\lambda - (\xi^2+2)/36\}^{\frac{1}{2}} ,$$

where ξ is the unit normal P fractile defined by $\Phi(\xi) = P$.

For very accurate work, use:

$v^{***} = (k + \frac{2}{3} - \lambda + \frac{.02}{k+1})(1 + A)^{\frac{1}{2}} \lambda^{-\frac{1}{2}}$, where

$A = (1 - f^2 + 2f \ln f)(1 - f)^{-2}$ and

$f = (k + \frac{1}{2})\lambda^{-1}$; Peizer and Pratt [7] give a table to determine A from f;

Never use $w = (k+\frac{1}{2} - \lambda)\lambda^{-\frac{1}{2}}$, $w_1 = (k-\lambda)\lambda^{-\frac{1}{2}}$ or $y = (k+1-\lambda)(k+1)^{-\frac{1}{2}}$, because our "quick work" suggestion is more accurate and not more laborious. However, w is useful as a first step in the calculation of v^{*} or v^{**} mentioned above.

Table 2

Advice for the approximation to confidence bounds for the Poisson parameter λ, when k events have been observed. In this table ξ denotes the unit normal $1-\alpha$ fractile, e.g., 1.9600 for $\alpha = 0.025$. The formulae give the lower bound with confidence coefficient $1-\alpha$; for the analogous upper bound replace k by k+1 and ξ by $-\xi$ throughout. The lower bound is zero for k = 0 and $-\ln(1-\alpha)$ for k = 1, the upper bound is $-\ln \alpha$ for k = 0.

For quick work, use:

$(k^{\frac{1}{2}} - \frac{1}{2}\xi)^2$ (but use $k - \xi k^{\frac{1}{2}}$ for $\alpha > 0.1$ and its upper bound analogue for $\alpha \geq 0.1$)

For more accurate work, use:

$\{(k+B)^{\frac{1}{2}} - \frac{1}{2}\xi\}^2$ for $\alpha \geq 0.025$, $k - \xi k^{\frac{1}{2}} + R$ otherwise, where $B = (\xi^2-4)/12$ and $R = (\xi^2-1)/3$; cf. Table 4;

or, still better,

$k\{1 - \frac{1}{3}\xi k^{-\frac{1}{2}} - \frac{1}{9}k^{-1}\}^3$.

For very accurate work, use:

$k - \xi k^{\frac{1}{2}} + R - Sk^{-\frac{1}{2}} - Tk^{-1}$, where $R = (\xi^2-1)/3$, $S = (\xi^3-7\xi)/36$ and $T = (3\xi^4+7\xi^2-16)/810$; cf. Table 4;

the alternative formula:

$\{(k + B + CE^{-\frac{1}{2}})^{\frac{1}{2}} - \frac{1}{2}\xi\}^2$, where $B = (\xi^2-4)/12$ and $C = (\xi^3+2\xi)/72$, cf. Table 4, is more accurate for lower bounds, with $E = \frac{3}{4}(k-1)$, if $\alpha \leq 0.05$ and roughly $k \leq 15$, for upper bounds, with $E = \frac{7}{6}k + 3$, if $0.01 \leq \alpha \leq 0.05$ and roughly $k \leq 25$.

Table 3

Advice for the approximation to the P fractile of the Poisson distribution with expectation λ. This fractile k satisfies $F_\lambda(k) = P$, and ξ denotes the unit normal P fractile, i.e., $\Phi(\xi) = P$. As the approximation (but also the exact fractile) is usually not an integer, the result will be rounded off in most applications.

For quick work, use:

$$(\lambda^{\frac{1}{2}} + \tfrac{1}{2}\xi)^2 - 1 .$$

For more accurate work, use:

$$\lambda + \xi\lambda^{\frac{1}{2}} + 2B \text{ if } 0.025 \le P \le 0.975,\ \{(\lambda-B)^{\frac{1}{2}} + \tfrac{1}{2}\xi\}^2-1 \text{ otherwise,}$$

where $B = (\xi^2-4)/12$; cf. Table 4.

For very accurate work, use:

$$\lambda + \xi\lambda^{\frac{1}{2}} + 2B - C\lambda^{-\frac{1}{2}} + T\lambda^{-1},$$

where $B = (\xi^2-4)/12$, $C = (\xi^3+2\xi)/72$ and $T = (3\xi^4+7\xi^2-16)/810$; cf. Table 4.

Table 4

Some functions of the unit normal P fractile ξ used in the preceding tables. Note that the odd polynomials C and S change sign with ξ, but the others are even.

$\alpha = 1 - P$	0.1	0.05	0.025	0.01	0.005
$p = 1 - \alpha$	0.9	0.95	0.975	0.99	0.995
ξ	1.281552	1.644854	1.959964	2.326348	2.575829
$B = (\xi^2-4)/12$	-0.1965	-0.1079	-0.0132	0.1177	0.2196
$C = (\xi^3+2\xi)/72$	0.0648	0.1075	0.1590	0.2395	0.3089
$R = (\xi^2-1)/3$	0.2141	0.5685	0.9472	1.4706	1.8783
$S = (\xi^3-7\xi)/36$	-0.1907	-0.1962	-0.1720	-0.1026	-0.0261
$T = (3\xi^4+7\xi^2-16)/810$	0.0044	0.0307	0.0681	0.1355	0.2006

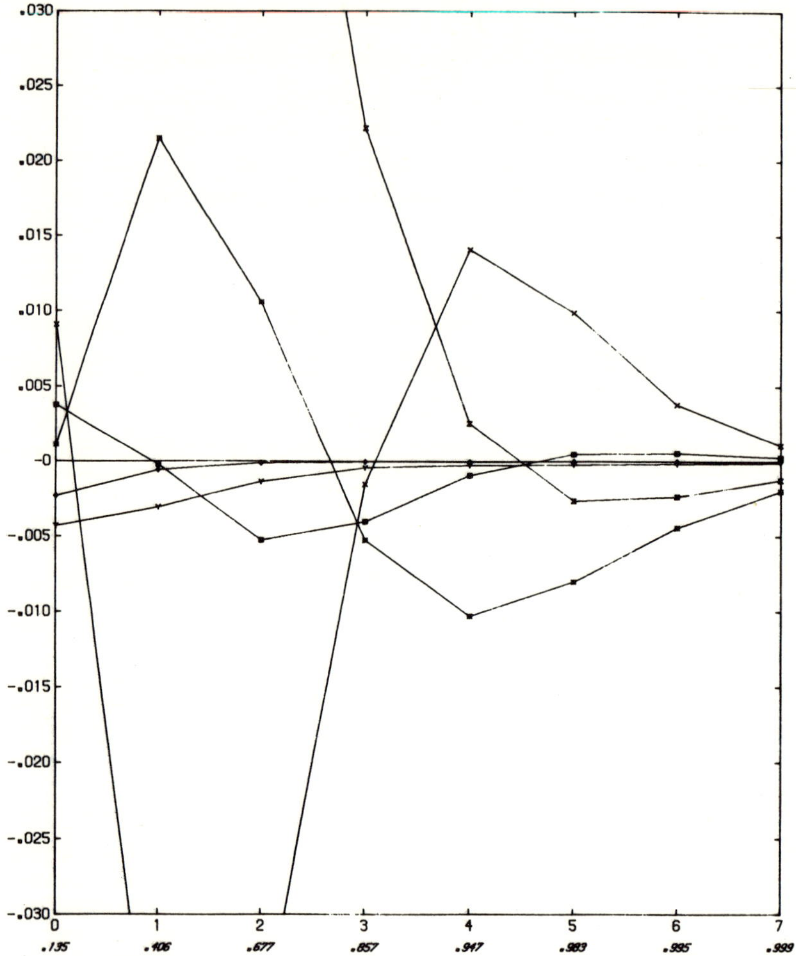

Fig. 1. Errors $\Phi(u) - F_\lambda(k)$ for $\lambda = 2$.
The horizontal scale gives k, and in italics the Poisson distribution function $F_\lambda(k)$. The vertical scale gives the errors. For the normal deviate u six functions of k and λ are considered. In the notation of Table 1 they are marked by × for w, ⧖ for v, ∗ for v_1, ◻ for v^{**}, Y for v^{**}, ◇ for v^{***}.

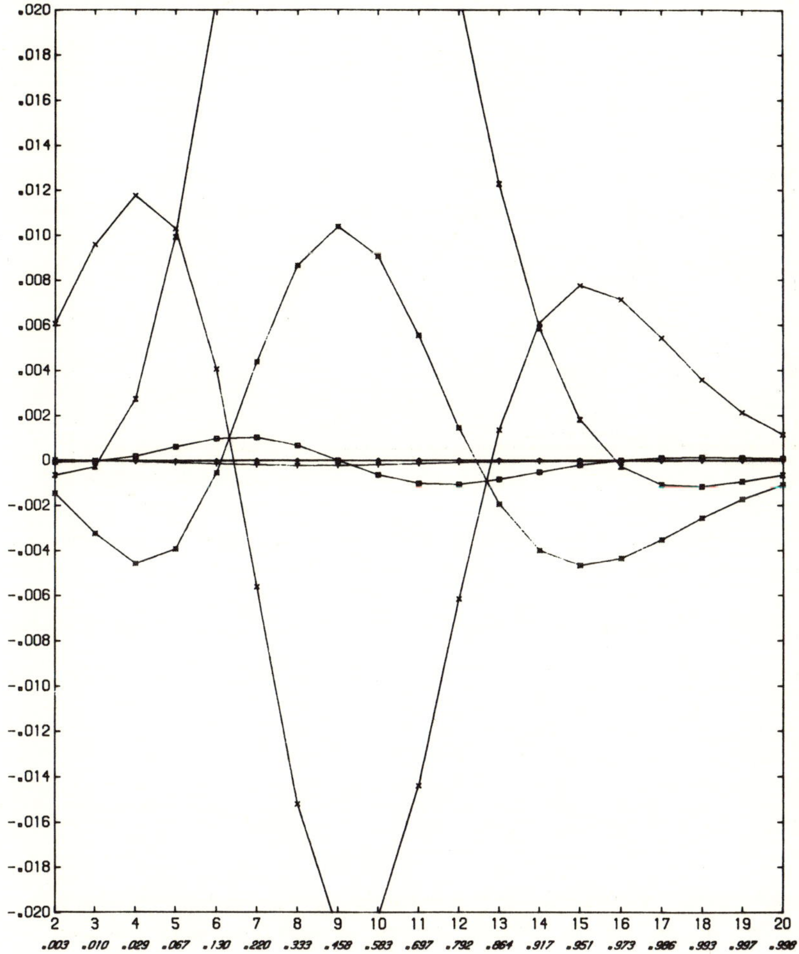

Fig. 2. Errors $\Phi(u) - F_\lambda(k)$ for $\lambda = 10$.
The horizontal scale gives k, and in italics the Poisson distribution function $F_\lambda(k)$. The vertical scale gives the errors. For the normal deviate u six functions of k and λ are considered. In the notation of Table 1 they are marked by × for w, ⧖ for v, * for v_1, ⊡ for v^*, Y for v^{**}, ◇ for v^{***}.

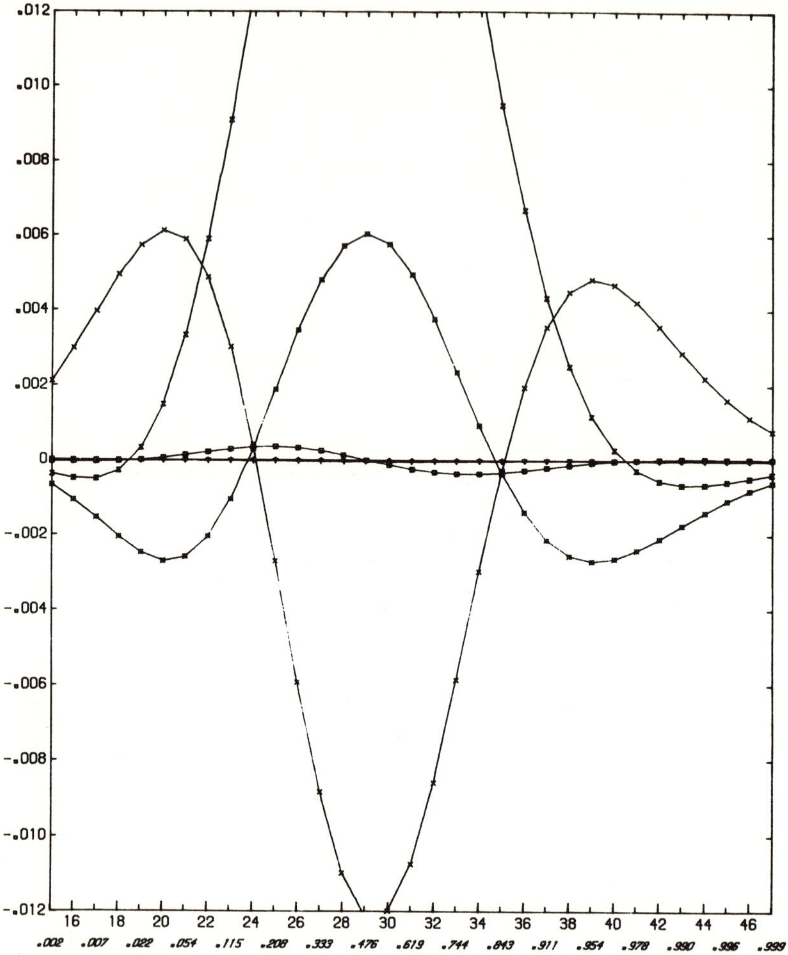

Fig. 3. Errors $\Phi(u) - F_\lambda(k)$ for $\lambda = 30$.
The horizontal scale gives k, and in italics the Poisson distribution function F (k). The vertical scale gives the errors. For the normal deviate u six functions of k and λ are considered. In the notation of Table 1 they are marked by × for w, ⧖ for v, ⋇ for v_1, ⊡ for v^*, Y for v^{**}, ◇ for v^{***}.

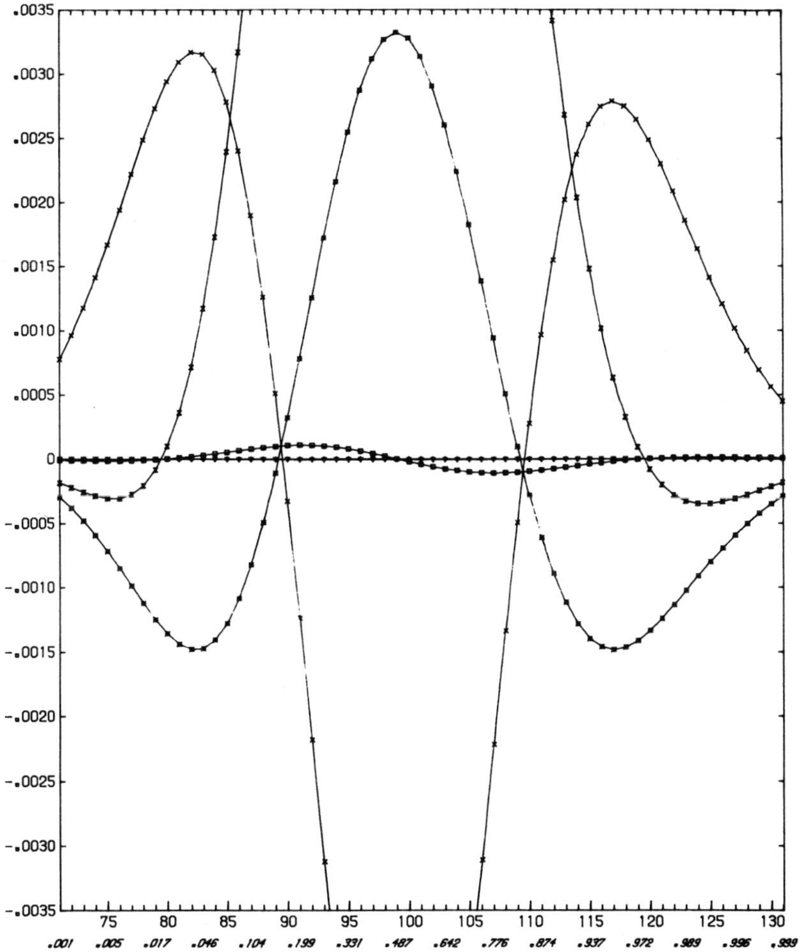

Fig. 4. Errors $\Phi(u) - F_\lambda(k)$ for $\lambda = 100$.
The horizontal scale gives k, and in italics the Poisson distribution function F (k). The vertical scale gives the errors. For the normal deviate u six functions of k and λ are considered. In the notation of Table 1 they are marked by × for w, X for v, ⋇ for v_1, ⊡ for v^*, Y for v^{**}, ◇ for v^{***}.

REFERENCES

[1] Anderson, T. W., and Burstein, H. 1967. Approximating the upper binomial confidence limit. J. Amer. Stat. Assoc. 62:857-61.

[2] ———, and ———. 1968. Approximating the lower binomial confidence limit. J. Amer. Stat. Assoc. 63:1413-16.

[3] Campbell, G. A. 1923. Probability curves showing Poisson's exponential summation. Bell System. Tech. J. 2:95-113; by G. A. Campbell, 1937, American Telephone and Telegraph Co., New York, 224-42.

[4] Cornish, E. A., and Fisher, R. A. 1937. Moments and cumulants in the specification of distributions. Revue de l'Institut International de Stat. 5:307-20.

[5] Esseen, C. G. 1945. Fourier analysis of distribution functions. A mathematical study of the Laplace-Gaussian law. Acta Math. 77:1-125.

[6] Goldberg, H. A., and Levine, H. 1946. Approximate formulas for the percentage points and normalization of t and χ^2. Ann. Math. Stat. 17:216-25.

[7] Peizer, D. B., and Pratt, J. W. 1968. A normal approximation for binomial, beta and other common, related tail probabilities. J. Amer. Stat. Assoc. 63:1417-56 and 1457-83.

[8] Riordan, J. 1949. Inversion formulas in normal variable mapping. Ann. Math. Stat. 20:417-25.

ON COCHRAN'S Q-TEST AND ITS MODIFICATION

V. P. BHAPKAR
University of Kentucky
Department of Statistics

SUMMARY

For the comparison of percentages in c matched samples [4] offers an exact conditional test based on the statistic Q, using randomization, and also a large-sample χ^2 approximation for the conditional distribution of Q. Alternatively, we have a large-sample unconditional test (offered by Bhapkar [2]), which is applicable also to the general case for the comparison of proportions in d categories ($d \geq 2$) of c matched samples. This paper illustrates the use of the latter statistic with some discussion on the relative merits of the two statistics and the hypotheses they are supposed to test.

1. INTRODUCTION

Suppose that each member of a sample is matched with a corresponding member in every other sample and that there are, thus, c matched samples. These c matched outcomes may be looked upon as a c-dimensional response whose components are generally correlated because of matching. Assume that each component of this response belongs to one of d possible categories ($d \geq 2$) and that there are r such independent observations, each with c components (or outcomes). Then the data consist of frequencies $n_{i_1,i_2,\ldots,i_c}$; the number of observations with response $i_1,i_2,\ldots i_c$, $i_j = 1,2,\ldots,d$ for each $j = 1,2,\ldots,c$. Thus, for convenience, we have numbered the categories $1,2,\ldots,d$. The sum over any suffix is indicated by writing zero in the place of that suffix.

For the special case $d = 2$, one may identify category 1 with "success" and 2 with "failure," so that we have

$n_{1,0,\ldots,o} = T_1$, $n_{0,1,0,\ldots,0} = T_2,\ldots,n_{0,\ldots,0,1} = T_c$, where T_j is the number of successes in the j-th sample, in the notation of Cochran [4]. To test the null hypothesis, say H_0, that for each observation every one of the c samples is considered equally likely to obtain a success, given that there are u_s successes for the s-th observation, $s = 1,2,\ldots,r$, Cochran [4] suggested the statistic

$$Q = \frac{c(c-1)\Sigma_j(T_j-\bar{T})^2}{c(\Sigma_s u_s)-(\Sigma_s u_s^2)} ; \tag{1}$$

the test proposed by him is a conditional test with u's regarded fixed. It involves computation of Q for every one of the $\pi_s \binom{c}{u_s}$ possible ways in which u_s successes can be distributed among the c samples for the s-th observation and the product is over only those s for which u_s is different from zero and c. This computation might be prohibitive even for samples of moderately large size; hence a χ^2 approximation with c-1 D.F. was suggested by him for Q to test H_0. For the case $c = 2$, Q reduces to the statistic given by McNemar [6].

One may approach the problem in an unconditional set-up and assume that

$p_j(i)$ = the probability that the response belongs to the i-th category for the j-th sample;

the natural null hypothesis to test is, then,

$$H_0^* : p_1(i) = p_2(i) = \cdots = p_c(i), \quad i = 1,2,\ldots,d. \tag{2}$$

Notice that in (2) we may restrict to $i = 1,2,\ldots,d-1$. For the special case $d = 2$, we may drop i and consider

$$H_0^* : p_1 = p_2 = \cdots = p_c, \tag{3}$$

where p_j is the probability of success for the j-th sample. To test H_0^* given by (3) the statistic proposed by Bhapkar [2] is

$$Q^* = \sum_{j=1}^{c} \sum_{k=1}^{c} a_{j,k} T_j T_k - \frac{(\Sigma_j T_j \Sigma_k a_{j,k})^2}{\Sigma\Sigma_{jk}\, a_{j,k}}, \tag{4}$$

where

$$[a_{j,k}]_{c \times c} = \left[T_{j,k} - \frac{T_j T_k}{r} \right]^{-1}, \tag{5}$$

T_j being the number of successes in the j-th sample, $T_{j,j} = T_j$ and $T_{j,k}$ ($j \neq k$) the number of observations with successes for the j-th and k-th components of the response; thus $T_{1,2} = n_{1,1,0,\ldots,0}, \ldots, T_{(c-1),c} = n_{0,\ldots,0,1,1}$. It was proposed to use Q^* as a χ^2 criterion with (c-1) D.F. for r large or even moderately large.

It should be pointed out here that the above Q^* with c-1 D.F. is applicable only if the quantities in (3), viz. p_j's are independent, in which case the matrix on the right hand side of (5) has an inverse. If this matrix is of rank v ($\leq c$), then deleting c-v dependent rows and columns of this matrix one can obtain Q^* with v-1 D.F.[1] for testing H_0^* after deleting the corresponding p's. The matrix whose inverse is required in (5) may be termed the sample covariance matrix of $[T_1, T_2, \ldots, T_c]$; its rank is v if and only if precisely v components of response are linearly independent in the given data.

---

[1] Alternatively, one may prefer a conservative test regarding c-1 as the D.F. of Q^*, even when the sample covariance matrix is of rank v.

2. EXAMPLE

The following example is taken from Siegel [10] to illustrate the use of Q* statistic. The artificial data are supposed to have come from an experiment to test the influence of interviewer friendliness upon housewives' responses in an opinion survey using 3 matched samples with 18 individuals. Here we have

$$c = 3, \quad r = 18,$$

$$T_1 = 13, \quad T_2 = 13, \quad T_3 = 3$$

$$T_{1,2} = 11, \quad T_{1,3} = 3, \quad T_{2,3} = 3.$$

Hence from (5)

$$[a_{j,k}] = \begin{bmatrix} 13 - \frac{13^2}{18} & 11 - \frac{13\times 13}{18} & 3 - \frac{13\times 3}{18} \\ 11 - \frac{13\times 13}{18} & 13 - \frac{13^2}{18} & 3 - \frac{13\times 3}{18} \\ 3 - \frac{3\times 13}{18} & 3 - \frac{3\times 13}{18} & 3 - \frac{3^2}{18} \end{bmatrix}^{-1}$$

$$= \frac{1}{7560} \begin{bmatrix} 2700 & -1080 & -540 \\ -1080 & 2700 & -540 \\ -540 & -540 & 3384 \end{bmatrix}$$

Therefore,

$$\Sigma_k \, a_{1,k} = \frac{1080}{7560}, \quad \Sigma_k \, a_{2,k} = \frac{1080}{7560}, \quad \Sigma_k \, a_{3,k} = \frac{2304}{7560},$$

and

$$\Sigma\Sigma \, a_{j,k} = \frac{4464}{7560}.$$

Thus, from (4),

$$Q^* = \frac{1}{7560} \left[493776 - \frac{(34992)^2}{4464} \right] = 29.03, \quad \text{D.F.} = 2.$$

Q* is highly significant; it may be pointed out here that Q = 16.7, which is also highly significant.

3. DISCUSSION

It may be worthwhile at this stage to look closely into the nature of these tests, the hypotheses they are supposed to test, and their relative advantages and disadvantages.

Cochran's Q statistic is based on $\Sigma_j(T_j - \bar{T})^2$, which is the same as the criterion used in the ordinary (i.e., Pearson) χ^2 test with independent samples; as such its basis is purely heuristic, and, as pointed out by Cochran himself, Q may not be the best criterion. As against this, Q* is a Wald's statistic [11] adapted to the problem of testing H_0^* and, as such, possesses some asymptotic optimal properties; in view of the identity of Neyman's χ_1^2 statistic [7], whenever it is defined, and Wald's statistic as proved by Bhapkar [3], Q* may also be looked upon as the Neyman χ_1^2 statistic for testing H_0^*. So from this point of view, Q* is expected to be more powerful than Q; this is reflected in the above example, with Q* much larger than than Q.

Q has the advantage of affording an exact test if one is prepared to compute Q for every possible arrangement of successes subject to the total number u_s for the s-th observation. This labor is worthwhile only for relatively small samples. The exact nature of the Q test is lost as soon as the χ^2 approximation is used and, in that case, both Q and Q* stand on the same footing as far as exactness is concerned.

Used as an exact conditional test, Q is supposed to test the hypothesis H_0 that, for each observation, every one of the c samples is considered equally likely to obtain a success. This chance of success, then, could conceivably vary from observation to observation, so that H_0^* appears to be unnecessarily more restrictive; but in such a situation the conclusions reached on the basis of Q should also be restricted only to the given observations and not in general. When the χ^2 approximation is used, the Q test is valid only for at least moderately large r, in which case the underlying hypothesis is usually taken to be H_0^* instead of H_0 (see, for example, Siegel [1956]).

It appears to me that, with the χ^2 approximation, the Q test is essentially used as an unconditional test, and hence, it is worthwhile examining the conditions under which Q has a limiting χ^2 distribution. It has been proved in the Appendix that under some conditions Q has a limiting χ^2 distribution with c-1 D.F.

only under H_0^{**} that the probability of success is the same for all samples and that the probability of simultaneous successes is also equal for every pair of samples; thus Q, as an unconditional statistic, is valid for testing

$$H_0^{**} : p_1 = p_2 = \cdots = p_c \text{ and } p_{1,2} = p_{1,3} = \cdots = p_{(c-1)c}, \quad (6)$$

where $p_{i,j}$ is the probability of simultaneous successes for the i-th and j-th samples, and not for just H_0^* alone. Note that H_0^* and H_0^{**} are equivalent only for the case of two samples; for $c > 2$, H_0^* and H_0^{**} are two different hypotheses, the latter being more restrictive than the former in the sense that H_0^{**} implies H_0^*. Q* is valid for H_0^* while Q, used unconditionally, is valid for H_0^{**}; in other words, Q is valid for testing H_0^* only under further restrictions. For testing H_0^{**}, however, one can construct an asymptotic χ^2 criterion given by the Wald statistic with D.F. = $(c-1) + [c(c-1)/2] - 1$, i.e., $(c^2+c-4)/2$, by the method given in [3].

Thus in the example discussed earlier with three matched samples, Q* is observed to be 29.03 on 2 D.F. for testing

$$H_0^* : p_1 = p_2 = p_3, \quad (7)$$

while Q is observed to be 16.7 on 2 D.F. for testing H_0^* under the further condition

$$C : p_{1,2} = p_{1,3} = p_{2,3}, \quad (8)$$

if used as an unconditional criterion. For testing both H_0^* and C, i.e.,

$$H_0^{**} : p_1 = p_2 = p_3, p_{1,2} = p_{1,3} = p_{2,3}, \quad (9)$$

one can, however, construct a suitable large-sample χ^2 criterion on 4 D.F.

Let us look more closely at the nature of H_0^{**} for the case of three samples. Suppose $p_{i,j,k}$ denotes the probability of response (i,j,k) in the earlier notation; in other words, $p_{i,j,k}$ is the probability that the first component of response belongs to the i-th category, the second component to the j-th category, and the third to the k-th category, i,j,k=1,2. With the convention that zero in place of a suffix indicates the sum over that suffix, we have

$$p_{i,j,0} = p_{i,j,1} + p_{i,j,2}, \quad \text{etc.}$$

$$p_{i,0,0} = p_{i,1,0} + p_{i,2,0}, \quad \text{etc.}$$

$$p_1 \equiv p_{1,0,0}, \; p_2 \equiv p_{0,1,0}, \; p_3 \equiv p_{0,0,1}$$

and

$$p_{1,2} \equiv p_{1,1,0}, \; p_{1,3} \equiv p_{1,0,1}, \; p_{2,3} \equiv p_{0,1,1}.$$

Thus H_0^* can be alternatively expressed as

$$H_0^* : p_{1,0,0} = p_{0,1,0} = p_{0,0,1} \tag{10}$$

and C as

$$C : p_{1,1,0} = p_{1,0,1} = p_{0,1,1}. \tag{11}$$

It is quite straight forward to establish that C is equivalent to, that is, C implies and is implied by

$$C^* : p_{1,1,2} = p_{1,2,1} = p_{2,1,1}. \tag{12}$$

Similarly, it is easy to see that H_0^* and C imply

$$D : p_{2,2,0} = p_{2,0,2} = p_{0,2,2}, \tag{13}$$

while C and D imply H_0^*; thus H_0^* and C are equivalent to C and D. But D is equivalent to

$$D^* : p_{1,2,2} = p_{2,1,2} = p_{2,2,1}, \tag{14}$$

while H_0^* and C is H_0^{**}. Thus H_0^{**} may be equivalently expressed as C* and D*, and

$$H_0^{**} : p_{1,1,2} = p_{1,2,1} = p_{2,1,1}, \; p_{1,2,2} = p_{2,1,2} = p_{2,2,1} \; . \tag{15}$$

In this representation H_0^{**} may be termed the hypothesis of complete symmetry of three components of response and, as noted earlier, is much stronger than just

$$H_0^* : p_{1,0,0} = p_{0,1,0} = p_{0,0,1},$$

which is the hypothesis of equality of marginal distributions. An asymptotic χ^2 criterion using Pearson-χ^2 based on the maximum likelihood estimates has been given by Sathe [9] to test such hypotheses of complete symmetry. For the case of three samples with just two categories for each response component, it reduces to

$$X^2 = \sum_{i,j,k}^{\{1,1,2\}} \frac{(n_{i,j,k} - n_*)^2}{n_*} + \sum_{i,j,k}^{\{1,2,2\}} \frac{(n_{i,j,k} - n_{**})^2}{n_{**}}, \qquad \text{D.F.} = 4 \tag{16}$$

where the first sum is over (i,j,k) taking value 1,1,2 in some order and, similarly, the second sum is over (i,j,k) taking values 1,2,2, in some order; also

$$n_* = (n_{1,1,2} + n_{1,2,1} + n_{2,1,1})/3$$

$$n_{**} = (n_{1,2,2} + n_{2,1,2} + n_{2,2,1})/3.$$

In the above example (Siegel [10], p. 164),

$$n_{1,1,2} = 8,\ n_{1,2,1} = 0,\ n_{2,1,1} = 0,$$

$$n_{1,2,2} = 2,\ n_{2,2,1} = 0,\ n_{2,1,2} = 2,$$

so that

$$n_* = \frac{8}{3} \text{ and } n_{**} = \frac{4}{3}.$$

Hence

$$X^2 = \frac{(8 - \frac{8}{3})^2 + (0 - \frac{8}{3})^2 + (0 - \frac{8}{3})^2}{8/3} + \frac{(2 - \frac{4}{3})^2 + (0 - \frac{4}{3})^2 + (2 - \frac{4}{3})^2}{4/3}$$

$$= 18.0,\ 4 \text{ D.F.}$$

which is highly significant.

Alternatively one may compute the Wald statistic for H_0^{**}; in view of the results in the earlier papers (Bhapkar [1] [3]), this asymptotic χ^2 criterion, Y^2, to test H_0^{**}, expressed in an equivalent form as

$$H_0^{**} : p_{1,0,0} - p_{0,1,0} = p_{1,0,0} - p_{0,0,1} = p_{1,1,0} - p_{1,0,1}$$

$$= p_{1,1,0} - p_{0,1,1} = 0,$$

is $\underset{\sim}{e}'\ \underset{\sim}{G}^{-1}\underset{\sim}{e}$ on 4 D.F., where e's are the expressions in H_0^{**} obtained after replacing p's by the corresponding sample proportions and G is their sample covariance matrix. For the above example it turns out that $\underset{\sim}{G}$ is of rank 3, so that one ultimately gets

$$Y^2 = 36.0,\ \text{D.F.} = 3,$$

for testing H_0^{**}; the computation details are omitted.

NOTE: One may still regard conservatively 4 as the D.F. of Y^2.

4. EXTENSION TO THE CASE d > 2

In the earlier notation, if there are possibly more than two categories for every component of the response, the natural hypothesis to test is H_0^* given by (2). It was shown by Bhapkar [2] that the Wald statistic Q* for the general case $d \geq 2$ is given by

$$Q^* = \sum_{j=1}^{c} \sum_{k=1}^{c} \underset{\sim}{t}_j' \underset{\sim}{A}_{j,k} \underset{\sim}{t}_k - \underset{\sim}{a}' \underset{\sim}{A}_0^{-1} \underset{\sim}{a}, \tag{17}$$

where

$$\begin{bmatrix} \underset{\sim}{A}_{1,1} & \cdots & \underset{\sim}{A}_{1,c} \\ \cdot & \cdot & \cdot \\ \underset{\sim}{A}_{c,1} & \cdots & \underset{\sim}{A}_{c,c} \end{bmatrix} = \begin{bmatrix} \underset{\sim}{T}_{1,1} - \frac{1}{r} \underset{\sim}{t}_1 \underset{\sim}{t}_1' & \cdots & \underset{\sim}{T}_{1,c} - \frac{1}{r} \underset{\sim}{t}_1 \underset{\sim}{t}_c' \\ \cdot & \cdot & \cdot \\ \underset{\sim}{T}_{c,1} - \frac{1}{r} \underset{\sim}{t}_c \underset{\sim}{t}_1' & \cdots & \underset{\sim}{T}_{c,c} - \frac{1}{r} \underset{\sim}{t}_c \underset{\sim}{t}_c' \end{bmatrix}^{-1},$$

with

$$\underset{\sim}{t}_1' = [n_{1,0,\ldots,0},\ n_{2,0,\ldots,0},\ldots,n_{(d-1)0,\ldots,0}],$$

$$\underset{\sim}{t}_c' = [n_{0,\ldots,0,1},\ n_{0,\ldots,0,2},\ldots,n_{0,\ldots,0,(d-1)}],$$

$$\underset{\sim}{T}_{1,1} = \text{diagonal } (n_{i_1,0,\ldots,0},\ i_1 = 1,2,\ldots,d-1), \text{ etc.,}$$

$$\underset{\sim}{T}_{1,2} = (n_{i_1,i_2,0,\ldots,0},\ i_1,i_2 = 1,2,\ldots,d-1), \text{ etc.,}$$

and

$$\underset{\sim}{A}_k = \Sigma_j\ \underset{\sim}{A}_{j,k},\ \underset{\sim}{A}_0 = \Sigma_k\ \underset{\sim}{A}_k,\ \underset{\sim}{a} = \Sigma_k\ \underset{\sim}{A}_k \underset{\sim}{t}_k.$$

Q* was proposed as a χ^2 criterion with (c-1) (d-1) D.F. for r large or even moderately large.

APPENDIX

As mentioned in the discussion, we prove here that under some conditions Q, unconditionally, has a limiting χ^2 distribution with c-1 D.F. as $r \to \infty$ under H_0^* only if (6) holds.

Let $X_{s,j} = 1$ if the j-th component of the response of the s-th individual is a success, and zero otherwise. Then

$$\varepsilon(X_{s,j}) = p_j,\ V(X_{s,j}) = p_j(1 - p_j),$$

$$\text{Cov}(X_{s,j}, X_{s,k}) = p_{j,k} - p_j p_k,\ j \neq k,$$

for $s = 1,2,\ldots,r$ and $j,k = 1,2,\ldots,c$; here $p_{j,k}$ is the probability of success for the j-th and k-th components simultaneously.

Note that $T_j = \Sigma_s X_{s,j}$. Let

$$\underset{\sim}{t}' = [T_1,\ldots,T_c] \text{ and } \underset{\sim}{p}' = [p_1,\ldots,p_c].$$

Then by the multi-dimensional form of the Central Limit Theorem (see, for example, Cramer [5], page 286) it follows that, as $r \to \infty$,

$$\frac{1}{\sqrt{r}}(\underset{\sim}{t} - r\underset{\sim}{p})$$

has limiting c-variate normal distribution with a null mean-vector and covariance-matrix

$$\underset{\sim}{H} = [h_{j,k}],$$

where

$$h_{j,k} = p_{j,k} - p_j p_k,$$

with the understanding that $p_{j,j} = p_j$. Note that under H_0^*, $p_1 = p_2 = \cdots = p_c = p$, say, so that

$$\underset{\sim}{p} = p\,\underset{\sim}{1},$$

where $\underset{\sim}{1}' = [1]$, a c-vector; also

$$\underset{\sim}{H} = \underset{\sim}{P} - p^2\,\underset{\sim}{L}, \tag{18}$$

where

$$\underset{\sim}{P} = [p_{j,k}],$$

with $p_{j,j} = p$ and $\underset{\sim}{L} = [1]$, a c×c matrix. Thus under H_0^*

$$\underset{\sim}{y} = \frac{1}{\sqrt{r}}(\underset{\sim}{t} - r\,p\,\underset{\sim}{1})$$

is asymptotically $N(\underset{\sim}{0},\underset{\sim}{H})$ with $\underset{\sim}{H}$ given by (18). Now

$$\frac{1}{r}\Sigma_j(T_j - \bar{T})^2 = \underset{\sim}{y}'\underset{\sim}{E}\,\underset{\sim}{y},$$

where

$$\underset{\sim}{E} = \underset{\sim}{I} - \frac{1}{c}\underset{\sim}{L}.$$

Also u_s in (1) is equal to $\Sigma_j X_{s,j}$ so that $\Sigma_s u_s = \Sigma_j T_j$ and

$$\Sigma_s u_s^2 = \Sigma_s [\Sigma_j X_{s,j}^2 + \sum_{j \neq k}\sum X_{s,j} X_{s,k}] = \Sigma_j T_j + \sum_{j \neq k}\sum T_{j,k} .$$

Therefore

$$\frac{1}{r} [c\Sigma_s u_s - \Sigma_s u_s^2] = \frac{1}{r} [(c-1)\Sigma_j T_j - \sum_{j \neq k}\sum T_{j,k}],$$

which, under H_0^*, converges in probability to

$$c(c-1)p - \sum_{j \neq k}\sum p_{j,k},$$

as $r \to \infty$ in view of Khinchin's Theorem (see, for example, Cramer [5], page 253). Hence in the limit as $r \to \infty$, Q in (1) is distributed as

$$\frac{1}{\sigma^2} \underset{\sim}{y}' \underset{\sim}{E}\, \underset{\sim}{y}$$

under H_0^*, where

$$\sigma^2 = p - \frac{1}{c(c-1)} \sum_{j \neq k}\sum p_{j,k} . \tag{19}$$

Since $\underset{\sim}{y}$ is asymptotically $N(\underset{\sim}{0},\underset{\sim}{H})$, if $\underset{\sim}{H}$ is nonsingular, Q is distributed as a χ^2 in the limit if and only if

$$\frac{1}{\sigma^2} \underset{\sim}{E}\, \underset{\sim}{H}\, \underset{\sim}{E} = \underset{\sim}{E} . \tag{20}$$

This follows from the well-known theorem (see, for example, Rao [8], page 152) concerning the necessary and sufficient condition for a quadratic form in a normal variate to have a χ^2 distribution.

Now

$$\underset{\sim}{E}\, \underset{\sim}{H}\, \underset{\sim}{E} = (\underset{\sim}{I} - \frac{1}{c} \underset{\sim}{L})(\underset{\sim}{P} - p^2\underset{\sim}{L})(\underset{\sim}{I} - \frac{1}{c} \underset{\sim}{L})$$

$$= \underset{\sim}{P} - \frac{1}{c} \underset{\sim}{P}\, \underset{\sim}{L} - \frac{1}{c} \underset{\sim}{L}\, \underset{\sim}{P} + \frac{1}{c^2} \underset{\sim}{L}\, \underset{\sim}{P}\, \underset{\sim}{L} ;$$

therefore the comparison of the j-th diagonal terms in the matrices in (20) gives

$$p - \frac{1}{c} \sum_{k=1}^{c} p_{j,k} - \frac{1}{c} \sum_{k=1}^{c} p_{j,k} + \frac{1}{c^2} \sum_{j=1}^{c} \sum_{k=1}^{c} p_{j,k} = (1 - \frac{1}{c})\sigma^2$$

so that

$$p - \frac{2}{c} (p + \sum_{k \neq j} p_{j,k}) + \frac{1}{c^2} \{cp + c(c-1)(p - \sigma^2)\} = \frac{c-1}{c} \sigma^2$$

in view of (19), which ultimately gives

$$\sum_{k \neq j} p_{j,k} = (c-1)(p - \sigma^2) \tag{21}$$

for every $j = 1,2,\ldots,c$. The comparison of jk-th terms in (20) further gives, for $j \neq k$,

$$p_{j,k} - \frac{2}{c}\{p + (c-1)(p-\sigma^2)\} + \frac{1}{c^2}\{cp + c(c-1)(p-\sigma^2)\} = \frac{-\sigma^2}{c}$$

in view of (21), finally reducing to

$$p_{j,k} = p - \sigma^2$$

for every jk, $j \neq k$. This establishes the assertion that, if $\underset{\sim}{H}$ is nonsingular, Q has in the limit χ^2 distribution with D.F. c-1 (which is the rank of $\underset{\sim}{E}$) only under H** given by (6).

REFERENCES

[1] Bhapkar, V. P. 1961. Some tests for categorical data. Annals of Mathematical Statistics 32:72-83.

[2] ———. 1965. Categorical data analogs of some multivariate tests. Institute of Statistics Mimeo series No. 450, 1965, University of North Carolina. To appear in the Roy Memorial Volume to be published by the University of North Carolina.

[3] ———. 1966. A note on the equivalence of two test criteria for hypotheses in categorical data. Journal of American Statistical Association 61:228-35.

[4] Cochran, W. G. 1950. The comparison of percentages in matched samples. Biometrika 37:256-66.

[5] Cramer, H. 1946. Mathematical Methods of Statistics. Princeton University Press.

[6] McNemar, Q. 1949. Psychological Statistics. John Wiley and Sons, New York.

[7] Neyman, J. 1949. Contribution to the theory of the χ^2 test. Proceedings of the Berkeley Symposium on Mathematical Statistics and Probability. Berkeley: University of California Press, 239-73.

[8] Rao, C. R. 1965. Linear Statistical Inference and its Applications. John Wiley and Sons, New York.

[9] Sathe, Y. S. 1962. Studies of some problems in nonparametric inference. Institute of Statistics Mimeo series No. 325, University of North Carolina.

[10] Siegel, S. 1956. Nonparametric Statistics for the Behavioral Sciences. McGraw Hill, New York.

[11] Wald, A. 1943. Tests of Statistical hypotheses concerning several parameters when the number of observations is large. Transactions of American Mathematical Society. 54:426-82.